AF686960

This work: "Teaching and Learning Mathematical Modelling: Approaches and Developments from German Speaking Countries", was originally published by Springer Nature and is being sold pursuant to a Creative Commons license permitting commercial use. All rights not granted by the work's license are retained by the author or authors.

Cover Image Credit/Copyright Attribution: IIerlok_Xolms/Shutterstock

Gilbert Greefrath · Katrin Vorhölter

Teaching and Learning Mathematical Modelling

Approaches and Developments from German Speaking Countries

Gilbert Greefrath
Institut für Didaktik der Mathematik und der Informatik
Westfälische Wilhelms-Universität Münster
Münster, Nordrhein-Westfalen
Germany

Katrin Vorhölter
Fakultät für Erziehungswissenschaft
Universität Hamburg
Hamburg, Hamburg
Germany

DOI 10.1007/978-3-319-45004-9

Library of Congress Control Number: 2016947918

© The Editor(s) (if applicable) and The Author(s) 2016. This book is published open access.
Open Access This book is distributed under the terms of the Creative Commons Attribution 4.0 International License (http://creativecommons.org/licenses/by/4.0/), which permits use, duplication, adaptation, distribution and reproduction in any medium or format, as long as you give appropriate credit to the original author(s) and the source, provide a link to the Creative Commons license and indicate if changes were made.
The images or other third party material in this book are included in the work's Creative Commons license, unless indicated otherwise in the credit line; if such material is not included in the work's Creative Commons license and the respective action is not permitted by statutory regulation, users will need to obtain permission from the license holder to duplicate, adapt or reproduce the material.
The use of general descriptive names, registered names, trademarks, service marks, etc. in this publication does not imply, even in the absence of a specific statement, that such names are exempt from the relevant protective laws and regulations and therefore free for general use.
The publisher, the authors and the editors are safe to assume that the advice and information in this book are believed to be true and accurate at the date of publication. Neither the publisher nor the authors or the editors give a warranty, express or implied, with respect to the material contained herein or for any errors or omissions that may have been made.

Main Topics You Can Find in This "ICME-13 Topical Survey"

- Development of modelling discussion in German-speaking countries
- Brief analysis of different modelling cycles and perspectives of modelling
- Mathematical modelling as a competency in the educational standards
- Role of technology in teaching and learning modelling
- Empirical research results on mathematical modelling from German-speaking countries.

Contents

Teaching and Learning Mathematical Modelling: Approaches and Developments from German Speaking Countries

1 Introduction

Mathematical modelling is a world-renowned field of research in mathematics education. The International Conference on the Teaching and Learning of Mathematical Modelling and Applications (ICTMA), for example, presents the current state of the international debate on mathematical modelling every two years. Contributions made at these conferences are published in Springer's *International Perspectives on the Teaching and Learning of Mathematical Modelling* series. In addition, the ICMI study *Modelling and Applications in Mathematics Education* (Blum et al. 2007) shows the international development in this area. German-speaking researchers have made important contributions in this field of research. The discussion of applications and modelling in education has a long history in German-speaking countries. There was a tradition of applied mathematics in German schools, which had a lasting influence on the later development and still has an impact on current projects. Two different approaches for different types of schools were brought together at the end of the last century. The relevance of applications and modelling has developed further since ICME 3, held in Karlsruhe in 1976.

In Germany, the focus on mathematical modelling has strongly intensified since the 1980s. Different modelling cycles were developed and discussed in order to describe modelling processes and goals as well as arguments for using applications and modelling in mathematics teaching. After subject-matter didactics (*Stoffdidaktik*[1]) affected mathematics education with pragmatic and specific approaches in Germany, there was a change in the last quarter of the 20th century towards a competence orientation, focusing on empirical studies and international cooperation.

[1]German words for some concepts are introduced in parentheses.

© The Author(s) 2016

G. Greefrath and K. Vorhölter, *Teaching and Learning Mathematical Modelling*, ICME-13 Topical Surveys, DOI 10.1007/978-3-319-45004-9_1

In 2006, Kaiser and Sriraman developed a classification of the historical and more recent perspectives on mathematical modelling in school. Mandatory educational standards for mathematics were introduced in Germany in 2003. Mathematical modelling is now one of the six general mathematical competencies. There have been many efforts for implementing mathematical modelling into school in Germany and modelling activities in mathematics teaching have changed in the last years due to the existence of digital tools.

Many recent qualitative and quantitative research studies on modelling in school focus on students; however, teachers also play an important role in implementing mathematical modelling successfully into mathematic lessons and in fostering students modelling competencies. In Germany there are now empirical studies on teacher competencies in modelling and other important topics. Furthermore, classroom settings play an important role. So apart from direct teacher behaviour, there has been a focus in research on the design of single modelling lessons as well as the whole modelling learning environment.

2 Survey on the State of the Art

2.1 Background of the German Modelling Discussion

The discussion of applications and modelling in education has played an important role in Germany for more than 100 years. The background of the German modelling discussion at the beginning of the 20th century differs between an approach of practical arithmetic (*Sachrechnen*) at the public schools (*Volksschule*, primary school and lower secondary school) and an approach supported by Klein and Lietzmann in the higher secondary school (*Gymnasium*).

In this context, arithmetic education evolved in the *Volksschule* in a completely different way than at the *Gymnasium* because there were initiatives requesting a stronger connection between arithmetic and social studies at the *Volksschule*. A book about teaching arithmetic at the *Volksschule*, *Der Rechenunterricht in der Volksschule*, written by Goltzsch and Theel in 1859, for example, outlines the importance of preparing students for their life after school. "Based on identical [mathematical] education, children should be prepared for the upcoming aspects of their life as well as for the manner in which numbers and fractions are widely applicable" (Hartmann 1913, p. 104, translated[2]). However, not everyone agreed on the importance of applications in mathematics education.

In the beginning of the 20th century, mathematics education was influenced by the reform pedagogy movement. Johannes Kühnel (1869–1928) was one of the representative figures in this movement. Kühnel demanded, that mathematics teaching to be more objective and interdisciplinary. Thus, arithmetic was supposed

[2]Unless otherwise noted, all translations are by the authors.

to become more useful and realistic. He considered the education of the 20th century to be very unrealistic. Distribution calculation, for example, included tasks where money had to be distributed in order to suit the specified circumstances. A characteristic example he gives is an alligation alternate problem that deals with a trader who has to deliver a certain amount of 60 % alcohol, but only has 40 and 70 % alcohol in stock. Students were asked to determine how many litres of each type should be mixed:

> To my great shame, I have to admit that in my whole life aside from school I never had to apply a distribution calculation, let alone an alligation alternate! I have never had to mix coffee or alcohol or gold or even calculate such a mixture, and hundreds of other teachers I interviewed admitted the same. (Kühnel 1916, p. 178, translated)

Above all, he criticised problems that involve an irrelevant context and demanded problems that were truly interesting for students. During these times, applications were considered to be more important for the learning process. They were used in order to help to visualise and motivate the students rather than prepare them for real life (Winter 1981). Apart from exercises dealing with arithmetic involving fractions and decimal fractions, there were commercial types of exercises referring to applied mathematics, such as proportional relations, average calculation, and decimal arithmetic. Kühnel's works were popular and widely accepted until the 1950s.

In contrast to the practical arithmetic approach at the *Volksschule*, the formal character of mathematics was in the centre of attention at the *Gymnasium*. Applications of mathematics were mostly neglected. This conflict was represented by two doctoral theses that were presented on the same day in Berlin. One was written by Carl Runge, later Professor of Applied Mathematics in Göttingen, the other one by Ferdinand Rudio, later Professor of Mathematics in Zürich (both cited after Ahrens 1904, p. 188):

- The value of the mathematical discipline has to be valued with respect to the applicability on empirical research (C. Runge, Doctoral thesis, Berlin June 23, 1880, translated).
- The value of the mathematical discipline cannot be measured with respect to the applicability on empirical research. (F. Rudio, Doctoral thesis, Berlin, June 23, 1880, translated).

Whereas Kühnel and other educators (representing the reform pedagogy movement) had a greater influence on the *Volksschule*, Klein started a reform process in the *Gymnasium*. In the beginning of the 20th century, a better balance between formal and material education was requested due to the impact of the so-called reform of Merano. The main focus was on functional thinking. In the context of the reform of Merano, a utilitarian principle was propagated "which was supposed to enhance our capability of dealing with real life with a mathematical way of thinking" (Klein 1907, p. 209, translated). Because of the industrial revolution, more scientists and engineers were needed. This is why applied mathematics gained in importance and real-life problems were used more often. Lietzmann (1919) makes important proposals for the implementation of Merano curricula and represents an implementation of applications in the classroom. Finally, the contents of the Merano reform in 1925 were nevertheless included in the curricula of Prussian secondary schools. The reform efforts were successful: "pragmatic objectives" were placed in the foreground of the curriculum from 1938 (Blum and Törner 1983).

This trend continued until the 1950s. In the late 1950s, Lietzmann stressed stronger inner-mathematical objectives (Kaiser-Meßmer 1986). After World War II, some ideas that had evolved from the progressive education movement and the reform of Merano were picked up again, but with applications losing importance. More emphasis was again placed on an orientation to the subject classification (Kaiser-Meßmer 1986).

New Math was a change in mathematics education during the 1960s and 1970s that aimed to teach abstract structures in mathematics to a higher degree. Surprisingly, applied mathematics did not vanish completely during these reforms, but it was influenced in different ways. Firstly, the mathematical core of a question was worked out more clearly, e.g., directly proportional and inversely proportional relationships. Secondly, the content of applications was extended, for example, by introducing probability at school, and, thirdly, methods were enhanced. For example, different visualisations by means of charts were discussed (Winter 1981). In the 1960s and 1970s, Breidenbach (1969) focused on the content structure of applications. He distinguished different levels of difficulty by the structural complexity of a question. Thus, he suggested ordering them accordingly. Comprehending the structure of a problem independently of its context and using the structure as a tool for students seems to be a convincing procedure. However, it is difficult for students to understand the entire structure of a problem before beginning to work on it. Studies show that students often switch between planning and processing while solving a problem (Borromeo Ferri 2011; Greefrath 2004). Hence, planning and implementation cannot be separated while dealing with complex problems. Furthermore, there is a risk of formalising mathematics education too strongly and thereby hindering students in finding their own creative ways to solve the problem (Franke and Ruwisch 2010). From the approach to solving word problems methodically, so-called arithmetical trees for students were developed, which visualise the structure of the word problem as a tree. These arithmetical trees still can be found in schoolbooks today. However, nowadays they serve the purpose of illustrating the structure of a calculation rather than revealing the structure of a word problem.

In the 1980s, the so-called New Practical Arithmetic (*Neues Sachrechnen*) evolved at all types of schools (Franke and Ruwisch 2010). The principles of the reform pedagogy movement were put in focus again and schools started to use applications in mathematics education more often. The New Practical Arithmetic aimed to find authentic topics for students and to carry out long-term projects that were supposed to be detached from the current mathematical topic and offer a variety of solutions. New types of questions, e.g., Fermi problems (Herget and Scholz 1998) were used accordingly. At the same time as the development of the

New Practical Arithmetic, the term *modelling* became better known in mathematics education (see Greefrath 2010). Initially, modelling was seen as a certain aspect of applied mathematics, which, to some extent, can be seen as an independent process within applications or as a perception of applications (Fischer and Malle 1985). In the 1980s and 1990s, Blum and Kaiser gradually introduced the term *modelling* into the German debate.

2.2 The Development from ICME 3 (1976) to ICME 13 (2016) in Germany

In 1976, Pollak gave a talk at ICME 3 in Karlsruhe, where he contributed to defining the term *modelling*. He pointed out that at that time it was less known how applications were used in mathematics teaching. To clarify the term, he distinguished four definitions of applied mathematics (Pollak 1977):

- Classical applied mathematics (classical branches of analysis, parts of analysis that apply to physics)
- Mathematics with significant practical applications (statistics, linear algebra, computer science, analysis)
- One-time modelling (the modelling cycle is only passed through once)
- Modelling (the modelling cycle is repeated several times).

There are distinct differences between these four definitions of applied mathematics. The first two definitions refer to the content (classical or applicable mathematics), whereas the other two relate to the processing procedure. Therefore, the term *modelling* focuses on the processing procedure. All four definitions are illustrated in a figure by Pollak (Fig. 1).

Modelling then was considered to be a cycle between reality and mathematics, which is repeated several times (Greefrath 2010).

To prepare the ICME-3 conference, Werner Blum, the coordinator for Section “B6, The Interaction Between Mathematics and Other School Subjects (Including Integrated Courses)”, undertook intensive research on the literature on mathematical modelling. Two volumes of documentation of selected literature on application-oriented mathematics instruction (Kaiser et al. 1982; Kaiser-Meßmer et al. 1992) resulted from this work later on. They provided an excellent overview of the national and international debate on applied mathematics education and also took into account selected publications on modelling that were written up to the beginning of the 20th century. The classification of works presented there incorporated ideas regarding goals, types of application, relation to reality, and embedding of the curriculum and analysed selected publications on applied mathematics teaching in more depth than ever before.

The classification system was presented at the First International Conference on the Teaching of Mathematical Modelling in 1983 in Exeter and had a significant

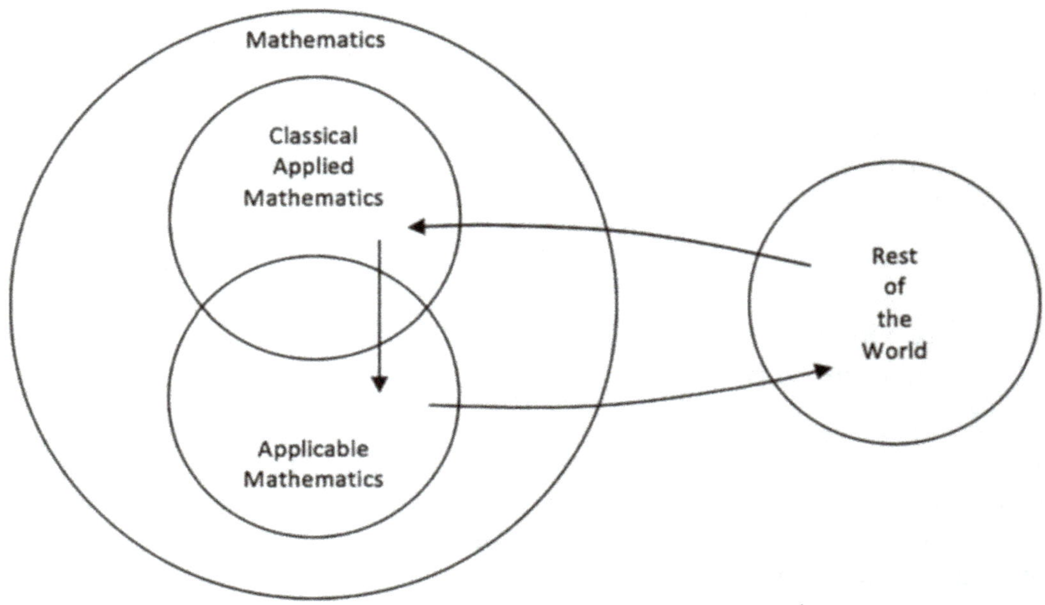

Fig. 1 Perspectives on applied mathematics by Pollak (1977, p. 256)

impact on a closer integration of German researchers, especially Werner Blum and Gabriele Kaiser, into the international debate on modelling (Blum and Kaiser 1984).

Henn (1980) gave an example of using mathematical modelling at school. He proposed the study of the theory of the rainbow as a piece of mathematics fraught with relations. This contribution was a revised version of his lecture delivered in 1979 in Freiburg at the German mathematics education conference. Many aspects of the rainbow were examined here and a mathematical model was presented. The model used an incident light beam and rays of first to fourth order. In addition, a detailed analytical model of a rainbow was developed. Thus, the occurring intensities could be described in detail. Furthermore, a model illustrating the reflected first-order ray was presented using dynamic geometry software. Thus it became obvious that explanations written in schoolbooks often contain mistakes.

The article of Blum (1985) about application-oriented mathematics instruction was very important in the modelling discussion in German-speaking countries. It included a range of application examples with a variety of topics, e.g., allocation of seats after elections, route mapping of motorway junctions, production of footballs, and granting of loans. Furthermore, this article showed that the debate on applications and modelling increasingly gained in importance. The best-known illustration of a modelling cycle in Germany (Fig. 2) can also be found in this contribution.

For the first time the visualisation shown in Fig. 2 is called a modelling process, which is based on the common concept at that time of models for mathematical application (Blum 1985, p. 200). Blum not only distinguished between applications and tasks, where the problem is wrapped into the context of another discipline or of

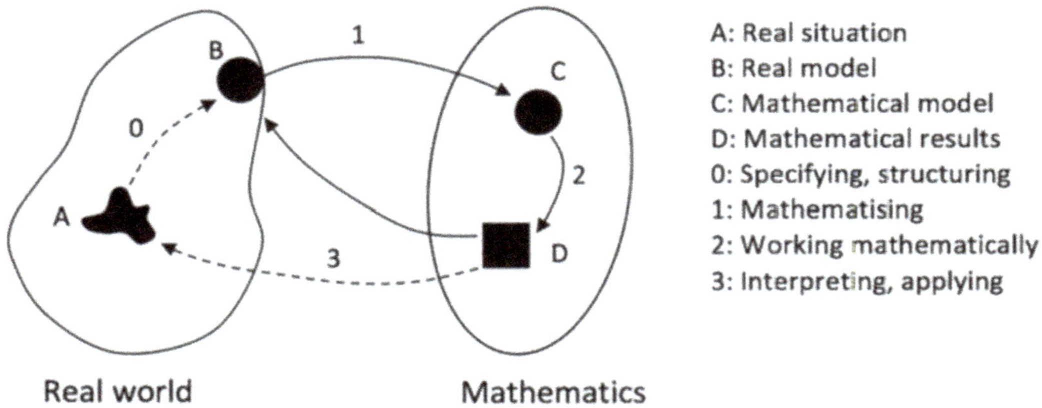

Fig. 2 Modelling cycle by Blum (Blum and Kirsch 1989, p. 134)

everyday life, he furthermore delivered arguments and aims regarding applications in mathematics teaching (i.e., objectives, arguments, and perspectives). In addition, he summarised arguments against applications such as time problems or less suitable examples. For details see Kaiser (2015).

In 1991, the German ISTRON Group was founded by Werner Blum and Gabriele Kaiser. This caused an intensified debate on modelling in Germany. The idea of ISTRON was that—for many reasons—mathematics education should put a greater focus on applications. Students should learn to understand environmental and real-life situations by means of mathematics and develop general mathematical skills (e.g., transfer between reality and mathematics) and attitudes such as open-mindedness regarding new situations. They should thereby establish an appropriate comprehension of mathematics including the actual use of mathematics. Learning mathematics should be supported by using relation to real life (Blum 1993).

A new series established in 1993 and published by Springer since 2014 enables the ISTRON Group, having already produced 20 volumes, to be present and visible in mathematics teaching as well as in the academic community. These contributions are intended to support teachers in dealing with real-life problems in school. Teachers are considered to be experts in teaching; therefore, teaching proposals should be modifiable so that teachers can adapt them to a specific situation. They should suggest uncommon ways of teaching mathematics and support preparing lessons (e.g., Bardy et al. 1996). In the following, some examples from the ISTRON volumes are presented.

The first volume of the ISTRON series resulted from a competition that was launched by the ISTRON Group at the end of 1991. They looked for contributions referring to teaching and learning mathematics that were combined with real-life applications, e.g., reports on teaching experience or new examples (see Blum 1993). The winning contribution of the international competition was also included in this volume: an article by Böer (1993) about a realistic extreme value problem. Böer explores the question of whether the packaging of one litre of milk with a square base, which was common at that time, was produced with a minimum of packaging material. The worksheet presented there is even today often used in mathematics education. Böer concluded that the optimal packaging of milk was only half a percent different from the real packaging used at that time (Greefrath et al. 2016).

The 14th International ICMI Study on Applications and Modelling in Mathematics Education Conference took place in 2004 in Dortmund, Germany. Werner Blum was the Chair of the IPC and Wolfgang Henn was the Chair of the Local Organising Committee. The accompanying ICMI Study volume fully presents the state of the discussion on modelling and applications at a high level. It became a standard reference work for the teaching and learning of applications and modelling. In addition, two conferences in the ICTMA series were held in Germany, the first in 1987 in Kassel (Blum et al. 1989) and the second in 2009 in Hamburg (Kaiser et al. 2015).

Over the following years mathematical modelling was incorporated into the curriculum and into the standards for mathematics education (see Sect. 2.8).

2.3 Mathematical Models

The debate about the term *mathematical model* plays an important role in the research on mathematical modelling in Germany. The term *modelling* describes the process of developing a model based on an application problem and using it to solve the problem (Griesel 2005). Therefore, mathematical modelling always originates from a real-life problem, which is then described by a mathematical model and solved using this model. The entire process is then called modelling.

As the development of a mathematical model as such is crucial, the term *mathematical model* shall be discussed in the following. A starting point for the definition of this term can be found in the publications of Heinrich Hertz. In the introduction of his book on the principles of mechanics, he described his considerations about mathematical models from a physical point of view. However, Hertz calls mathematical models "virtual images of physical Objects" (Hertz 1894, p. 1, translated). He mentions three criteria that should be used to select the appropriate mathematical model.

> Different virtual images are possible and they can even differentiate from various directions. Images not compatible with our commonly accepted rules of thinking should not be accepted. Therefore, all virtual images should be logically compliant or at least acceptable in the short term. Virtual images are false if their internal interdependencies are contradictory to the interdependencies of the external objects: they should be true. However, even two images both true and acceptable could differentiate in terms of expedience. Normally an image would be preferable that reflects more interdependencies than another, i.e., that is more concrete. If both images are equally compliant and concrete, the image of choice would be the least complex one. (Hertz 1894, p. 2f, translated)

Hertz mentions (logical) *admissibility*, *accuracy*, and *expediency* as criteria. A mathematical model is admissible if it does not contradict the principles of logical thinking. In this context, it is accurate if the relevant relations of a real-world problem are shown in the model. Finally, a model is expedient if it describes the matter by appropriate as well as relevant information. If a model proves to be expedient, it can only be judged in comparison with the real-life problem. It can be expressed by an economical model or in a different situation by the richness of relations (Neunzert and Rosenberger 1991). A new problem might require a new model, even if the object is the same. Furthermore, Hertz emphasises as *conditio sine qua non* that the mathematical model has to match the real-life items (Hertz 1894).

The term *mathematical model* has been described in the German literature in many ways. Models are simplified representations of the reality, i.e., only reflecting aspects being to some extent objective (Henn and Maaß 2003). For this purpose, the observed part of reality is isolated and its relations are controlled. The subsystems of these selected parts are substituted by known structures without destroying the overall structure (Ebenhöh 1990). Mathematical models are a special representation of the real world enabling the application of mathematical methods. If mathematical methods are used, mathematical models that just represent the real world can even deliver a mathematical result (Zais and Grund 1991). Thus, a mathematical model is a representation of the real world, which—although simplified—matches the original and allows the application of mathematics. However, the processing of a real problem with mathematical methods is limited, as the complexity of reality cannot be transferred completely into a mathematical model. This is usually not even desired. Another reason for generating models is the possibility of processing real data in a manageable way. Thus, only a selected part of reality will be transferred into mathematics through modelling (Henn 2002).

As it is often possible to simplify in different ways, models are not distinct. Because there are different types of models (see Fig. 3), it is even harder to describe the modelling process accurately. Prescriptive models are called normative models. Furthermore, models can be used as afterimages. These are called descriptive models (Freudenthal 1978). Characteristics of descriptive models are predictions and descriptions (Henn 2002).

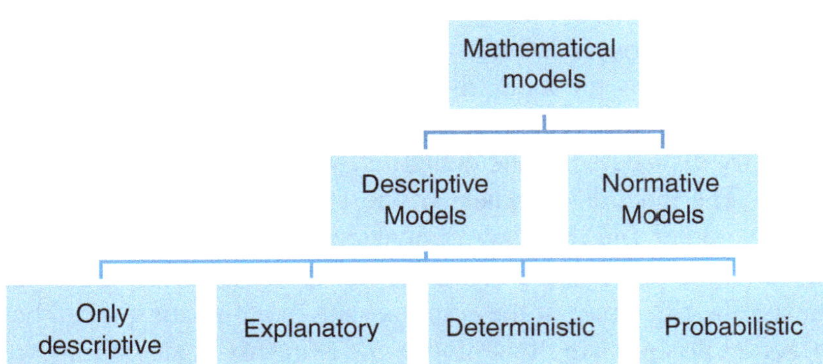

Fig. 3 Descriptive and normative models

Descriptive models aim to simulate and represent real life. This can happen in a descriptive or even explanatory way (Winter 1994, 2004). Therefore, one kind of descriptive model does not intend only to describe a selected part of reality but to help understanding the inner coherence. Furthermore, it is possible to distinguish between models aiming for understanding and models predicting a future development (Burscheid 1980). These predictions might be completely determined as well as to some extent probable. To summarise, there are descriptive models that are just descriptive in character, others that have additional explanations for something (explicative descriptive models), and, finally, those that even predict a development (deterministic and probabilistic models).

Tasks on descriptive and normative mathematical models can be quite different. Whereas descriptive models are used to describe and finally solve real-life problems, normative models aim to create mathematical rules as help in decision making in certain situations.

For example, to distribute the cost of heating in a house with several apartments, a normative model is needed. Actually, this is a real problem that students at the junior secondary level are able to understand and solve. Maaß (2007) offered a lesson plan regarding this problem, helping students to learn that different models can equally be a correct solution for the same problem. In this example, the reality was only created after deciding on a certain mathematical model, e.g., distribution of costs with respect to area, number of people, or consumption.

As modelling is characterised as a procedure for processing a problem, it can be seen as a difference between a conscious and an unconscious process. Reflection of the proceeding not being considered as a criterion for implementing mathematical modelling is called general perception. According to this general perception, a modelling process even occurs if it happens unconsciously (Fischer and Malle 1985). In the framework of this perception of modelling, students working on real-life problems without consciously simplifying the situation on a higher mathematical level are performing modelling.

2.4 Modelling Cycle

The entire modelling process is often represented as a cycle. The following is an easy example of outlining the modelling cycle. In order to calculate the volume of sand in a container, the problem must first be simplified by, for instance, assuming the sand is evenly distributed in the container, with the fill level roughly matching the loading sill. The material thickness of the container also need not be included, thus allowing the outer and the inner dimensions of the container to be equal. It is also reasonable to assume that the container has no bumps or other irregularities. In order to transfer the filled part of the container into mathematics, it can be identified with a trapezoidal prism. Using this model, the respective calculations will provide a mathematical solution. This solution can be interpreted as the volume of the sand (see Fig. 4).

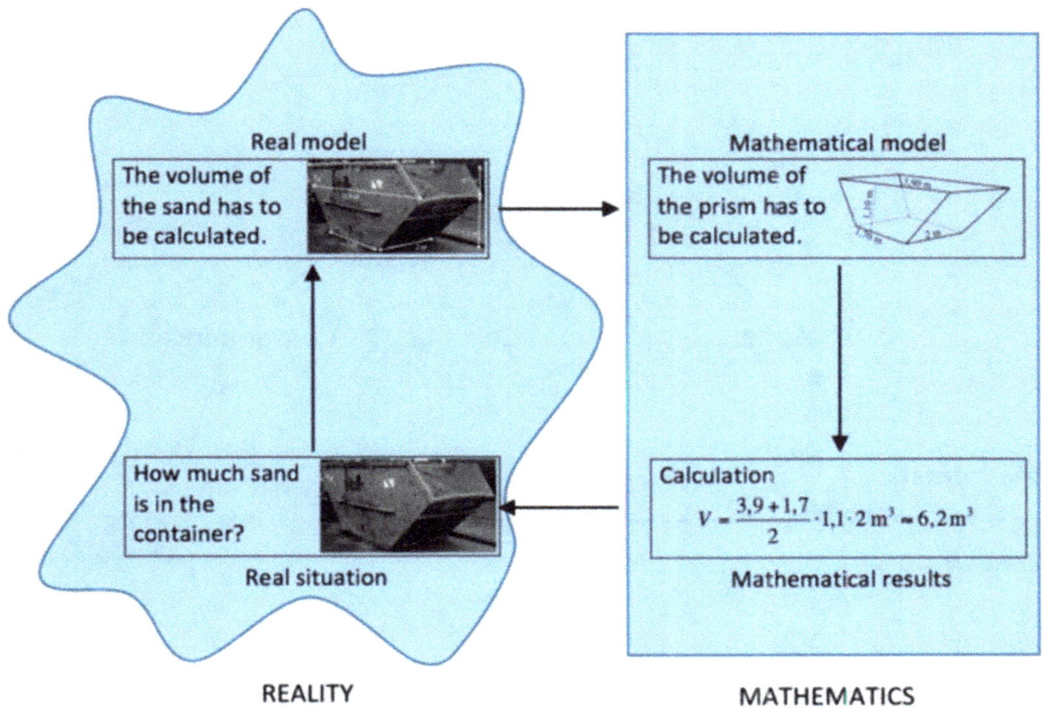

Fig. 4 Ideal problem-solving process of a problem shown as modelling cycle

The problem involving the volume of the sand in the container is a real-world problem. The first simplifications on a factual level lead to what is called a real-world model. Afterwards this is transferred to a mathematical model, which is used to calculate a mathematical solution. The result is then applied to the real-life problem.

It is also possible to idealise the solution process in other ways. For example, collecting the data could be shown separately or steps in developing the mathematical model could be omitted. Hence, different representations of the modelling cycle can be found in the literature. We present different descriptions of modelling processes in the following ordered by the complexity of steps in developing a mathematical model.

Single mathematising

If only one step is used to transfer a real-life problem to a model, this model of a modelling cycle is called single mathematising. In particular, the representation of the generally accepted model by Schupp (1988) is as clear as concrete. In one dimension, it divides mathematics and reality, which is common for models of mathematical modelling, while in the other dimension, the problem and solution are equally distinguished (see Fig. 5).

The modelling cycle need not always be fully completed or be repeated several times. Büchter and Leuders (2005) described the repeated modelling cycle as a spiral, i.e., emphasizing the evolution of experience over the modelling process.

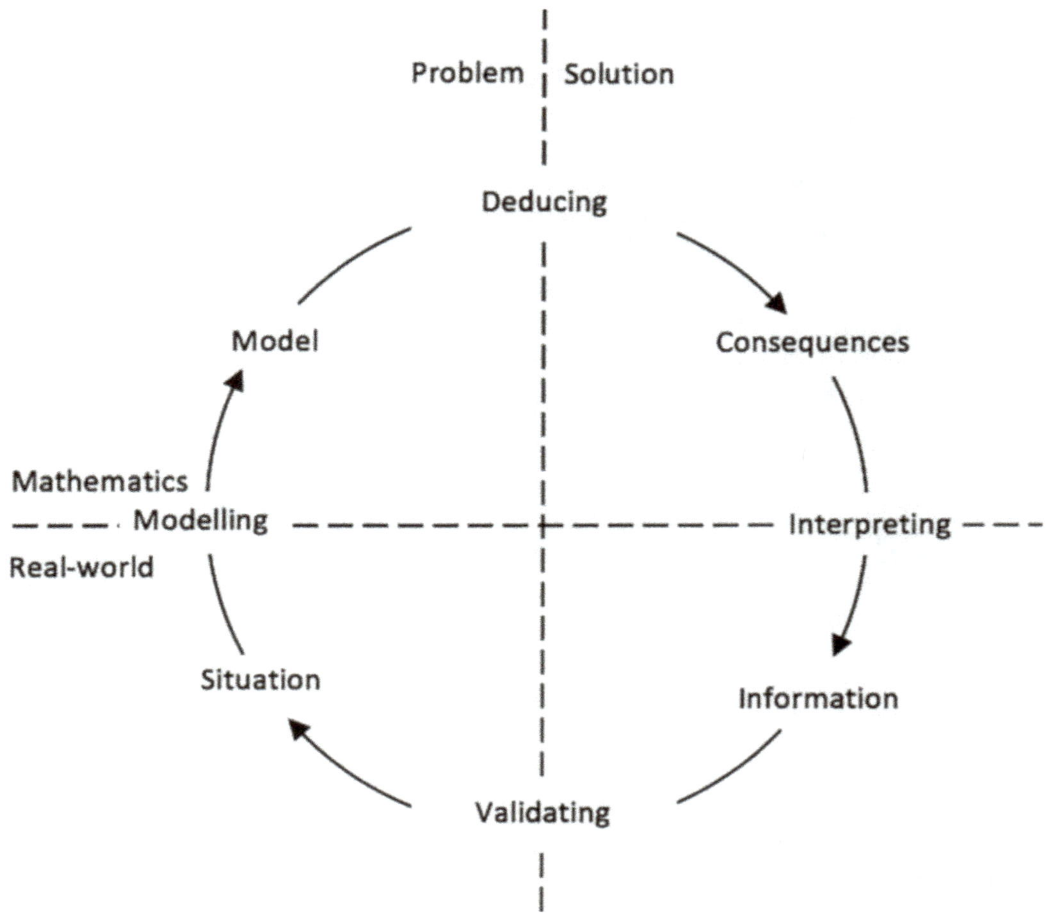

Fig. 5 Modelling cycle by Schupp (1989, p. 43)

After each run, experience with regard to solving the problem is gained. Büchter and Leuders also distinguished between real and mathematical models. However, specifying the problem is separated as an individual step between reality and model. There are also particular modelling cycles that include a simple mathematizing step.

The best-known modelling cycle in Germany was created by Blum (1985 see Fig. 2). It specified an additional step in building the mathematical model. Simplifying reality or, in other words, creating a real model was seen as an individual step (This has been used to solve the container problem shown in Fig. 4). This model was developed together with Kaiser-Meßmer (1986) and has been enhanced by many authors (e.g., Henn 1995; Humenberger and Reichel 1995; Maaß 2002; Borromeo Ferri 2004). In addition, Maaß (2005) as well as Kaiser and Stender (2013) added the interpreted solution as a step between mathematical solution and reality (see Figs. 6 and 7). This highlights interpreting and validating as different processes in the second half of the modelling cycle (see Greefrath 2010).

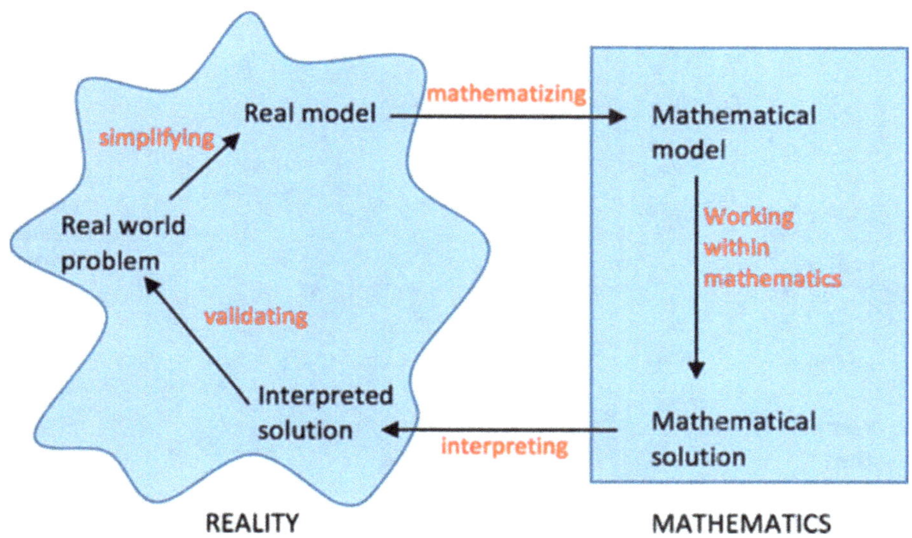

Fig. 6 Modelling cycle of Maaß (2006, p. 115)

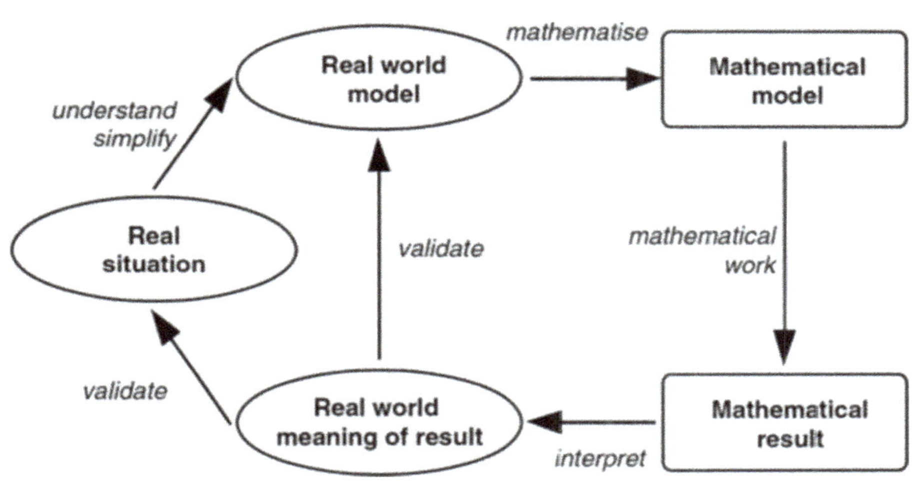

Fig. 7 Modelling cycle of Kaiser and Stender (2013, p. 279)

Complex mathematising

A newer model by Blum and Leiß (2005) and adapted by Borromeo Ferri (2006), was developed from a cognitive aspect (see Fig. 8). Blum's original model from 1985 was extended by the addition of a situation model, which showed more detail in considering how a mathematical model is generated. The role of the individual creating the model was also described in a more detailed way. The situation model outlined the individual's mental representation of the situation.

The model by Fischer and Malle (1985) described how to transfer a real-life situation to a mathematical model in detail. Interestingly enough, the process of collecting data was added to this model, which was specifically helpful in

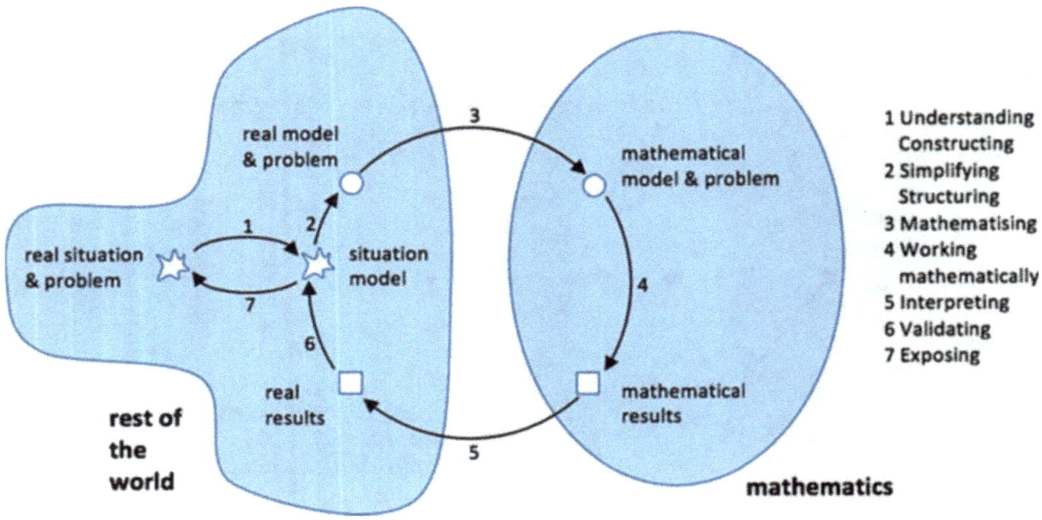

Fig. 8 Modelling cycle of Blum and Leiß (2005) (cited after Blum and Leiß 2007, p. 225)

specifying the simplification step. This description of the modelling process is especially suitable for Fermi problems, because most of the data have to be estimated.

Depending on target group, research topic, and research interest, the described models focus on different aspects. Often they also have a different purpose. Normative and descriptive models should especially be distinguished. For example, a certain model could be used to describe student activities within an empirical study. For this purpose, even very complex models are suitable (see Fig. 8). In a normative way, modelling cycles such as those shown in Fig. 5 could support students working on modelling problems in classes (see Greefrath 2010).

2.5 *Goals, Arguments, and Perspectives*

2.5.1 Goals

Different goals at various levels are pursued while using applications and modelling in mathematics teaching. Due to the link between mathematics and reality, mathematical modelling offers the unique opportunity to get interesting impressions in the subject of mathematics as well as in real life. Lietzmann (1919) already mentioned the goals for mathematics in this context, but also pointed out difficulties: "The application of mathematical facts to real life is of equal importance to the even heavier challenge of identifying mathematical problems in reality." However, he did not use the term *modelling*.

In what follows, content-related, process-oriented, and general goals of modelling are distinguished in order to underline the importance of mathematical modelling at different levels (see Blum 1996; Greefrath 2010; Kaiser-Meßmer 1986; as well as the overview by Niss et al. 2007).

Content-related goals
Content-related goals incorporate the pragmatic assumption that students working on modelling problems challenge their environment and are able to explore it by means of mathematics. The goal is—as it is for word problems related to modelling as a didactical direction—the ability to be aware of and understand phenomena of the real world. This corresponds to the first of three of what Winter (1996) called the *fundamental experiences*, which every student should get to know.

Process-oriented goals
In particular, interaction with applications in mathematics education requires general mathematical skills such as problem-solving capabilities. Essential heuristic strategies for problem solving, e.g., working with analogies or working with reverse calculation, can be used and encouraged in working on modelling problems. In addition, modelling problems particularly encourage communicating and arguing. This formal justification of modelling corresponds to Winter's third fundamental experience for a general mathematics education: "Mathematics education is fundamental because problem-solving capabilities far beyond mathematical tasks are learned." (1996, p. 37, translated). The goals of learning psychology also refer to the learning process. They focus on understanding and remembering mathematics by dealing with modelling. In the context of modelling, increasing motivation as well as general interest in mathematics is often named as a main goal.

General goals
Cultural arguments in particular have been mentioned as the most important general goals. Mathematics education should provide a balanced picture of mathematics as a science. The use of mathematics in the environment is crucial for the development of mathematics science and for democratic society. This also includes educating students to become responsible members of society who are able to critically judge models that are used daily, e.g., tax models. Social skills can also be taught by co-working on modelling problems (Greefrath et al. 2013).

2.5.2 Arguments

In the argumentation for applications there were originally only three goals for applied mathematics education. Blum (1985) divided them into four: Firstly, pragmatical arguments (i.e., mathematics as vehicle for special applications) should contribute to a better understanding of and coping with relevant extra-mathematical situations. Secondly, the use of applications for promoting general skills and attitudes, which cannot be helpful immediately for special relevant situations, was mentioned (called *formal arguments*). This new category was differentiated further: Methodological qualifications (meta-knowledge and general skills for applying mathematics) should be promoted. This can be done by getting to know general strategies for dealing with real situations by using examples. Especially in the translation between reality and mathematics, reflecting about applications and estimating the possibilities as well as the limits of applications in mathematics should be discussed. Furthermore, Blum subsumed the support of other general skills under these formal arguments. This entails the competence for arguing and problem solving as well as general attitudes towards openness to problem situations, which today is called general skills. Thirdly, Blum described the use of applications for giving the students an overall image of mathematics (arguments on the philosophy of science). In accordance with the third goal, applications are used for conveying a balanced impression of mathematics as a cultural and social phenomenon (Blum 1978). Fourthly, applications were seen as a help for learning mathematics (arguments on the psychology of learning). These corresponded to the second level of Blum (1978) and are divided into content-related aids (i.e., a local and a global structure of the content) and student-related support, which are intended to help improve understanding of mathematics and long-term retention of information as well as provide a better attitude towards mathematics (Blum 1985).

In addition, to differentiating the four arguments, which relate to modelling and application and contrast with the utilitarian view (this view aims to teach only the mathematics that is necessary for applications and modelling and the mathematical models that are bound to specific situations), the debate on mathematical modelling has been promoted significantly by emphasising meta-knowledge and general skills. For details, see Kaiser (2015).

2.5.3 Perspectives

Based on the analysis of the historical and current development of applications and modelling in mathematics education, different theoretical perspectives can be identified in the national and international debate on modelling. In her extensive analysis, Kaiser-Meßmer (1986) used three dimensions: a concept-related dimension referring to the importance of applications within the goals of mathematics education, a curricular dimension focussing on the role of applications in class, and a situational dimension taking the degree of reality of applications into account. At the beginning of the 21st century in the light of this analysis, Kaiser and Sriraman (2006) developed a classification of the historical and more recent perspectives on mathematical modelling in school. Different tendencies in the historical and current debate on applications and modelling can be distinguished, which are further differentiated in newer works on perspectives of modelling. In the German-speaking area, the following perspectives are particularly important.

Realistic and applied modelling
This tendency pursues content-related goals: solving realistic problems, understanding the real world, and encouraging modelling skills. It focuses on real and—above all—authentic problems in industry and science, which are only marginally

simplified. Modelling is seen as act where authentic problems are solved. The modelling process is not carried out in parts but as a whole. Real modelling processes, which are conducted by applied mathematicians, serve as role models. The theoretical background of this tendency is closely related to applied mathematics and historically relates to pragmatic approaches to modelling, which have been developed by Pollak (1968), among others, in the beginning of the newer modelling debate (see Kaiser 2005 as an example).

Pedagogical modelling
The purpose of this tendency includes process-related and content-related goals. It can be distinguished further into didactical and conceptual modelling.

Didactical modelling includes on the one hand encouraging the learning process of modelling and on the other hand dealing with modelling examples to introduce and practise new mathematical methods. Thus, modelling is completely incorporated into mathematics teaching.

The intent of conceptual modelling is to enhance students' development and understanding of terminology within mathematics and with regard to modelling processes. This also includes teaching meta-knowledge of modelling cycles and judging the appropriateness of the used models. The problems used for pedagogical modelling are developed for mathematics teaching in particular and are therefore simplified significantly (see Blum and Niss 1991; Maaß 2004 as examples).

Socio-critical modelling
Pedagogical goals and a critical understanding of the world are aimed at in order to critically examine the role of mathematical models and mathematics in general in society. The basic focus is not on the modelling process itself and its visualisation. Emancipatory perspectives on and socio-critical approaches to mathematics education are the background (see Gellert et al. 2001; Maaß 2007 as examples).

Cognitive modelling
This approach is seen as a kind of meta-perspective because it focuses on scientific goals. It is about analysing and understanding the cognitive procedures that happen in modelling problems. Hence, different descriptive models of modelling processes are developed, such as individual modelling paths for individual students. Psychological goals, e.g., supporting mathematical thinking in the light of cognitive psychology, also play a role. See Blum and Leiß (2005) and Borromeo Ferri (2011) as examples for this perspective (Greefrath et al. 2013).

2.6 *Classification of Modelling Problems*

Modelling processes can be specifically encouraged at school by means of adequate modelling problems. There is a broad range between short, less realistic questions that only focus on a partial competency and authentic modelling problems, which are worked on during a longer period of time (see Sect. 2.8). Modelling problems can be distinguished into a range of different problem categories (see for example Blum and Kaiser 1984; Greefrath 2010; Maaß 2010). The level of reality can be described more precisely using the categories authenticity, relevance to everyday life, realism, and relevance to students. Furthermore, assumptions in reality and in the task itself can be distinguished (Blum and Kaiser 1984). In their comprehensive documentation of relevant examples, Kaiser et al. (1982) distinguish the level of application: routine use of mathematical methods, reasonable application of mathematical methods depending on the situation and, furthermore, mathematisation of a situation and developing the terms and methods that are adequate for a model. In addition, both the level of reality (i.e., realistic versus consciously alienating reality) as well as the intention of a problem (i.e., mathematics helping to solve the problem versus using the problem to motivate and illustrate mathematical content) are analysed (see Kaiser et al. 1982; Blum and Kaiser 1984). In a comprehensive classification scheme, Maaß (2010) also takes into account which modelling activity supports the problem, which parts of the modelling process have to be done, what the type of context is, what the relation to reality is, what the level of openness in the question is, and what the cognitive requirements are.

2.7 *Modelling as a Competency and the German Educational Standards*

Based on results of the Danish KOM project (Niss 2003) and accompanied by international comparative studies, mandatory educational standards for mathematics were introduced in Germany beginning in 2003 (first in middle schools). Mathematical modelling is now one of the six general mathematical competencies that the education standards for mathematics rate as obligatory for intermediate school graduation. It can also be found in the education standards for primary school as well as for upper secondary school.

By means of different mathematical content, students are to acquire the ability to translate between reality and mathematics in both directions. In works of Blum (see Blum et al. 2007), modelling skills are described in a more detailed way as the ability to adequately perform the necessary steps in the process of changing back and forth between reality and mathematics as well as analysing and evaluating models in comparison.

The discussion of the different complex modelling cycles (see Sect. 2.2) showed that there exist various descriptions of modelling. These modelling cycles describe the different sub-processes of modelling with a different level of detail and emphasis. The ability to perform such a sub-process can be seen as a partial competence of modelling (Kaiser 2007; Maaß 2004). Looking at the modelling cycle in Fig. 8, these partial competences could be characterised as presented in Table 1. By means of detailed descriptions, also called indicators, the definition of

Table 1 Sub-competencies involved in modelling (see Greefrath et al. 2013, p. 19; Greefrath 2015)

Sub-competency	Indicator
Constructing	Students construct their own mental model from a given problem and thus formulate an understanding of their problem
Simplifying	Students identify relevant and irrelevant information from a real problem
Mathematising	Students translate specific, simplified real situations into mathematical models (e.g., terms, equations, figures, diagrams, and functions)
Interpreting	Students relate results obtained from manipulation within the model to the real situation and thus obtain real results
Validating	Students judge the real results obtained in terms of plausibility
Exposing	Students relate the results obtained in the situational model to the real situation, and thus obtain an answer to the problem

partial competences becomes obvious. Thus, an extensive list of partial modelling skills can be obtained. Working mathematically (students work with mathematical methods in the mathematical model and get mathematical solutions) is not listed as a partial competency, because it is not specific to the modelling process. By using different modelling cycles, other partial competences emphasising other aspects of modelling could occur.

It is possible to consciously divide modelling into partial processes to reduce the complexity for teachers and students and to create suitable exercises. This view of modelling especially enables training of individual partial competencies and establishing a comprehensive modelling competency in the long term. For more information on modelling competencies, refer to the comprehensive overview by Kaiser and Brand (2015).

The German educational standards for mathematics at the secondary level of 2003—as well as the educational standards at the primary level of 2004 and for higher education entrance qualification of 2012—describe mathematical modelling as a competency. The educational standards for the general higher education entrance qualification, for example, display the requirements regarding the modelling competency in the three following areas:

Requirement area I: Students can:

- Apply familiar and directly apparent models
- Transfer real situations directly into mathematical models
- Validate mathematical results with regard to the real situation.

Requirement area II: Students can:

- Carry out modelling processes consisting out of several steps and with few and not clearly formulated restrictions
- Interpret results of such modelling processes
- Adjust mathematical models to varying facts.

Requirement area III: Students can:

- Model complex real situations whereby variables and conditions have to be determined
- Check, compare, and evaluate mathematical models considering the real situation (KMK 2012, p. 17, translated).

Since 2006, an overall strategy regarding educational monitoring in Germany has been pursued by the Standing Conference of the Ministers of Education and Cultural Affairs. It aims to strengthen competence orientation within the educational system. The general competency in modelling plays an important role in mathematics. In addition to international school achievement studies (PISA, TIMSS), there are national achievement studies as well as comparative studies (VERA). These tests are carried out in classes in Grades 3 and 8 in all general education schools in order to investigate which competencies students have achieved at a particular point of time. The comparative studies aim to give teachers individual feedback on the educational standards requirements that students can handle.

Beginning in 2017, a pool with audit tasks for the *Abitur* examination will be provided for Germany from which all states can take audit tasks for the *Abitur*. This will be an important step in improving the quality of audit tasks and gradually adjusting the level of requirements in all states. Tasks are developed based on the educational standards; thus, by default some of the tasks for the *Abitur* include modelling as a competency.

2.8 Implementing Modelling in School

There have been many efforts to implement mathematical modelling into school in German-speaking countries: Besides collections of tasks [for example, the ISTRON series discussed in Sect. 2.2 and the collection of tasks by MUED (www.mued.de)], teaching unit for different goals (e.g., for fostering students modelling competencies as a whole, tasks with the same mathematical content, etc.) have been created in different projects aimed at fostering students modelling competencies in different ways. In addition, theoretical concepts for improving students' modelling competencies systematically and permanently have been developed by Böhm (2013).

Due to the high number of smaller and larger projects, we cannot present all of them. Therefore, we will focus on a special way of implementing modelling that has been initiated by several universities in various parts of Germany: modelling weeks or modelling days.

Modelling weeks or days were originally developed at the University of Kaiserslautern by the working group of Helmut Neunzert, have been carried out at the University of Hamburg for more than a decade, and have been adopted by universities such as Darmstadt, Munich, and Kassel. The structure of all modelling weeks or days is similar to those in Kaiserslautern or Hamburg. During modelling days and weeks, students of different ages (depending on the special project) are asked to work on a highly complex task for whole school days. Modelling weeks usually last one week and take place outside school (usually at a university or a youth hostel) while modelling days only last two or three days and take place in a school.

A central feature of these projects has been the use of highly complex modelling problems, often coming from research or industry. They have been simplified only slightly and normally introduced by a short presentation. Some of the problems that have been tackled so far have been:

- Pricing for Internet booking of flights
- Optimal automated irrigation of a garden
- Chlorination of a swimming pool
- Optimal distribution of bus stops.

Participating students have been asked to choose one of the offered tasks. Afterwards they were divided into different groups according to their interest.

The main purpose of modelling weeks and days has been to enable students to carry out modelling problems independently. Therefore, they have been supervised by tutors. In some cases, university teachers have supervised the students. In other cases, such as in Hamburg and Kassel, university students were trained to supervise.

The evaluation of modelling weeks and days has regularly shown great approval and learning outcome in various types of competencies (for more details see Kaiser and Schwarz 2010; Kaiser et al. 2013; Vorhölter et al. 2014).

2.9 Modelling and Digital Tools

Possible modelling activities in mathematics teaching have changed in the last years mainly due to the existence of digital tools. Especially when dealing with realistic problems, a computer or an adequately equipped graphical calculator can be a useful tool to support teachers and students. Henn (1998), for example, suggested this early on and proposed to implement digital tools, e.g., notebooks with algebra software, because this would enable the introduction of complex applications and modelling into daily teaching (see also Henn 2007).

Currently, digital tools are often used to work on such problems, e.g., to process models with complex function terms or to reduce the calculation effort. Digital tools can perform a range of tasks in teaching applications and modelling. One possibility for using these tools is experimenting and exploring (see Hischer 2002). For example, a real situation can be transferred to a geometrical model or it can be experimented on within this model by means of dynamic geometry software or a spreadsheet analysis. Very similar to experimenting is simulating. Simulations, which are experiments that use models, are intended to provide insights into the real system presented in the model or into the model itself (Greefrath and Weigand 2012). Predictions on the population of a certain animal species with different environmental conditions, for example, are possible by means of a simulation. Applied mathematics simulations done by computer can be understood as a part of a modelling cycle in which a numerical model that was developed from the mathematical model is tested and validated by comparing it with measurement results (Sonar 2001). Deterministic simulations with fixed problem data and stochastic simulations taking random effects into account are distinguished (Ziegenbalg et al. 2010).

A common use of digital tools, especially computer algebra systems, is calculating or estimating numerical or algebraic solutions (see Hischer 2002); without these tools students would not be able to make these estimations, at least within a reasonable time frame. A computer can also be used to do calculations to find algebraic representations from the information given. In addition, digital tools can perform a visualisation of a subject being taught in school (Barzel et al. 2005; Hischer 2002; Weigand and Weth 2002). For example, the data given can be represented in a coordinate system by means of a computer algebra system or a statistics application. This can be a starting point to develop mathematical models. Digital tools also play a useful role in controlling and verifying (Barzel et al. 2005). Therefore, digital tools can help with control processes for discrete functional models, for example. If computers with internet connection are provided for mathematics teaching, they can be used to do investigations (Barzel et al. 2005), e.g., in context with applications. In this way, real problems can be understood in the first place and simplified afterwards.

A computer's different capacities can be used in mathematics education for a range of steps in the modelling cycle. Control processes, for example, are usually the last step of the modelling cycle. Calculations are done by means of the generated mathematical model, which in analysis, for example, is often represented by a function. Some possibilities for using digital tools during the modelling process are represented in the modelling cycle in Fig. 9, which is modified from Blum and Leiß's modelling cycle (see Fig. 8). Digital tools can be usefully applied in every step of the modelling cycle.

If the steps in calculating with digital tools are looked at more precisely, working on modelling problems with digital tools requires two translation processes. First, the modelling question has to be understood, simplified, and translated into mathematics. The digital tool, however, can only be used after the mathematical terms have been translated into the computer's language. The results calculated by the computer then have to be transformed back again into mathematical language. Finally, the original problem can be solved when the mathematical results are applied to the real situation. These translation processes can be represented in an extended modelling cycle (see Fig. 10), which in addition to the rest of the world and mathematics also includes technology (see Greefrath and Mühlenfeld 2007; Savelsbergh et al. 2008; Greefrath 2011). Current studies, however, show that actual modelling activity that includes a computer can be better described by the integrated view.

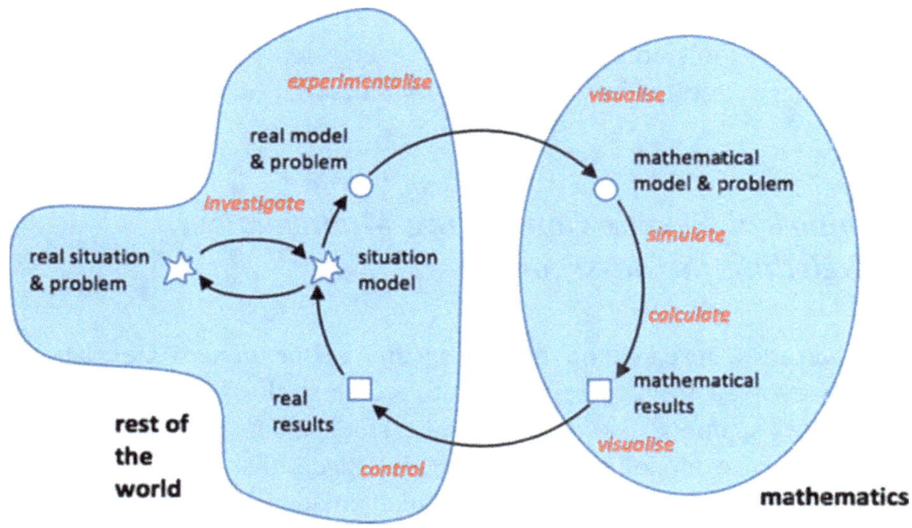

Fig. 9 Possible use of digital tools for modelling (Greefrath 2011, p. 303)

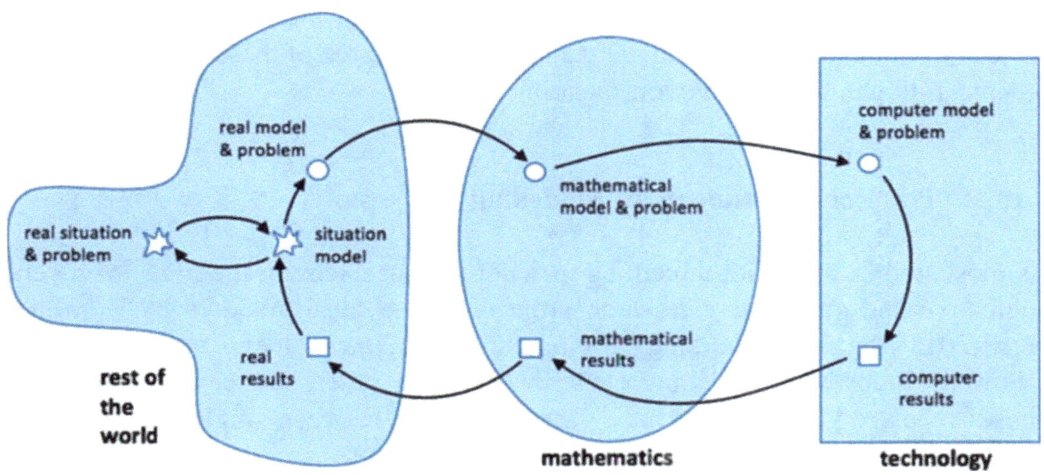

Fig. 10 Extended modelling cycle (Greefrath 2011, p. 302)

Currently, there exists little empirically established knowledge about the possibilities of teaching modelling and the limits of working with digital tools in mathematics teaching, as there have been case studies rather than large-scale implementation studies. Open research questions can be found in the works of Niss et al. (2007). These include the following questions: How are digital tools supposed to be used in different grades to support modelling processes? What is the effect of digital tools on the spectrum of modelling problems to be worked on? How is teaching culture influenced by the existence of digital tools? When do digital tools enhance or hinder learning opportunities in the modelling process?

Additional empirical research is required to clarify the questions named above, especially considering the extended modelling cycle and the necessary translation processes. The case studies by Greefrath et al. (2011) and Geiger (2011) point out though that digital tools could be useful for every step of the modelling process. This is particularly true for interpreting and validating.

2.10 Empirical Results Concerning Mathematical Modelling in Classrooms

In the last decades, research on modelling and application in German-speaking countries has evolved from merely qualitative case studies to larger research projects with bigger samples, also including case studies. The main focus was on the factors that influence modelling processes, on aspects that have to be considered while trying to implement modelling into mathematics lessons, and on possible ways to optimally improve students modelling competence. The studies therefore incorporated the actors of modelling processes: students of different ages and teachers, modelling problems, and learning settings. In the following, central research results from German-speaking countries from the last decades are presented. Although many of them are related to two or three of these aspects, they are ordered following the distinction mentioned above.

2.10.1 The Role of Students in Modelling Processes

As most studies deal with modelling in school, students are in the focus of several qualitative and quantitative research projects. It was clearly shown by Borromeo Ferri (2011) that when working on modelling problems, students normally do not follow the steps of a modelling cycle in the given order. Rather they pass some phases repeatedly and omit others. Often they skip between single phases, which are called "mini-loops" by Borromeo Ferri. Similar results concerning individuals dealing with modelling problems can be found in Leiß (2007) and Greefrath (2004).

The results of the early study of Maaß (2006) clearly showed "that modelling competencies include more competencies than just running through the steps of a modelling process." (Maaß 2006, p. 139). Thus, one of the important aspects is the connection between pure modelling competencies on the one hand and different kinds of competencies on the other hand. Working successfully and being goal oriented on modelling problems requires various competencies such as mathematical competencies, reading competencies, and metacognitive competencies. The influence of these competencies on the modelling process has been investigated in several projects. Furthermore, the interplay between different students' beliefs and preferences and students' modelling capabilities has been analysed. Selected results will be outlined in the following.

As one of the main influencing competencies, various studies have focused on *mathematical competency* as an indispensable competency in working on mathematical models. In different qualitative and quantitative studies, a strong relation between this sub-competence and modelling competence as a whole was verified. Important results concerning this relation can be found in Main Study 2 of the DISUM Project. In this classical intervention study with 21 classes in Grade 9, students' achievement and attitudes during a 10-lesson teaching unit were tested with the help of various tests and questionnaires. A correlation between mathematical competence and modelling competence was exposed (Leiß et al. 2010).

Also within the framework of Main Study 2 of the DISUM Project, comprehensive *reading competency* (i.e., reading texts as well as capturing tables and graphics) was identified as an important influencing factor. In order to analyse the connection between reading competence and modelling competence, two different kinds of reading tests were used: A general reading test and a mathematical reading test. The results showed that both reading tests measure the same theoretical construct. Furthermore, on the basis of the results of the study, mathematical reading competence was identified as a prerequisite for successful work on modelling tasks (Leiß et al. 2010).

Metacognitive competencies have been identified as a third influential factor on solving modelling problems. Both nationally and internationally, research on metacognition has evolved in educational psychology, general education, and mathematics education. In doing so, declarative meta-knowledge has been distinguished from procedural meta-knowledge (often called metacognitive strategies) (for further descriptions, see Vorhölter and Kaiser 2016). Qualitative and quantitative research in the last decades has focused on both aspects. For example, in her qualitative study, Maaß (2006) identified a relation between declarative meta-knowledge about the modelling process and modelling tasks on the one hand and modelling competencies on the other hand. Furthermore, she identified single weaknesses in modelling that match with certain misconceptions in meta-knowledge (Maaß 2006).

In a quantitatively oriented study, 86 ninth graders from 10 different classes were asked to report on their use of learning strategies and metacognitive strategies while solving modelling problems. In addition, their modelling competencies were tested. No significant correlation between cognitive and metacognitive self-reported strategies (in general or task orientated) on the one hand and mathematical modelling competence on the other hand were found. As one reason for this result, the measurement of metacognitive strategies was identified (Schukajlow and Leiß 2011). Therefore, Blum summarises: "One of the problems in these empirical studies is how to measure strategy knowledge, on the one hand, and strategy use, on the other hand, and another problem is how to reliably link students' activities to their strategies." (Blum 2015, p. 88).

In addition to these results on competencies, students' characteristics have been identified as another influencing factor on student performance while working on modelling problems. In the following, some of the main studies are briefly presented.

In a case study with 35 students, Maaß (2006) reconstructed four *types of modellers* on the basis of their attitude towards context and mathematics:

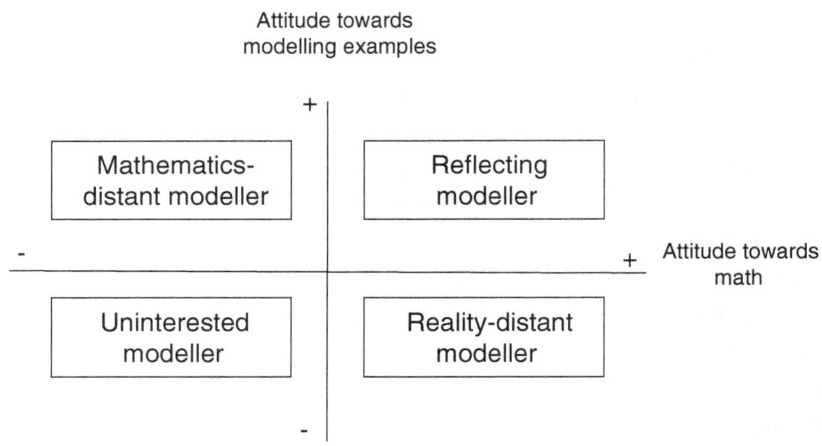

Fig. 11 Types of modellers (Maaß 2006, p. 138)

(1) reality-distant modeller, (2) mathematics-distant modeller, (3) reflecting modeller, and (4) uninterested modeller (see Fig. 11).

While reality-distant modellers are described as overwhelmingly positive towards mathematics without reference to the context, mathematics-distant modellers are characterised as preferring the context and being rather opposed to mathematics. According to this classification, reality-distant modellers have weaknesses in stages that require consulting reality. Mathematics-distant modellers on the other hand have deficits in working mathematically. Combinations of these two types are the reflecting modeller and the uninterested modeller. Whereas the uninterested modeller is interested neither in the context nor in mathematics, the reflecting modeller has a positive attitude both to the context and to mathematics. Thus, the reflecting modeller shows an appropriate performance while working on the problem, whereas the uninterested modeller shows deficits in all steps of the modelling process (Maaß 2006).

With the help of this classification, Maaß worked out the impact of attitudes on the development of modelling competencies: A negative attitude towards modelling tasks (i.e., uninterested modeller and reality-distant modeller) appeared to hinder the development of modelling performance, especially the development of sub-competencies necessary for setting up a real model and validating the solution. Those students performing well in mathematics were for the most part able to overcome existing weaknesses in the phases of setting up a real model and validating. On the contrary, students not performing well in mathematics were not able to doing so (Maaß 2006).

In a case study with 35 students, Borromeo Ferri (2010), referring to work by Burton (2004), identified *mathematical thinking style* as another influencing factor on student performance while working on modelling problems. Students with an analytic thinking style tend to switch very fast from the real situation to mathematics and focus on these mathematical phases. Students with a visual thinking style on the contrary merely begin to work on a problem by verbalising their mental model and building a real model. For students with an integrated thinking style, no typical procedure could be reconstructed (Borromeo Ferri 2010).

In a case study with 8 students aged 16–17, Busse (2005) reconstructed four types of dealing with the contextual aspects of a modelling problem. Two extremes, a reality-bound and a mathematics-bound type were distinguished: Students of the first type try to solve a task only by using non-mathematical concepts and methods. Students who are mathematics bound on the contrary perceive the context of a task merely as decoration. They translate necessary contextual information into mathematics at once and do not use further personal knowledge. As a combination of these two types, Busse identified an integrating type who uses the given information as well as personal knowledge in order to mathematise and solve the task and validate the solution. These students apply mathematical methods to solve the task. In contrast, representatives of the ambivalent type (which is a combination of the reality-bound and the mathematical-bound types) internally prefer contextually accentuated reasoning. Externally, they prefer a mathematical reasoning. Different from the integrating type, in the ambivalent type both ways do not complement each other but coexist (Busse 2005) (Fig. 12).

All these studies show that students' competencies and characteristics have a great influence on students' work on modelling problems. Some of the factors are necessary for solving modelling tasks successfully and influencing the individual approach, while some are obstructive. For promoting modelling competencies effectively, Blum therefore summarises: "It is important to care for a parallel development of competencies and appropriate beliefs and attitudes. Taking into account the remarkable stability of beliefs and attitudes, this also requires long-term learning processes" (Blum 2015, p. 86).

To summarise, students are no "blank pages" when starting to work on a modelling problem. On the contrary, different competencies, characteristics, and

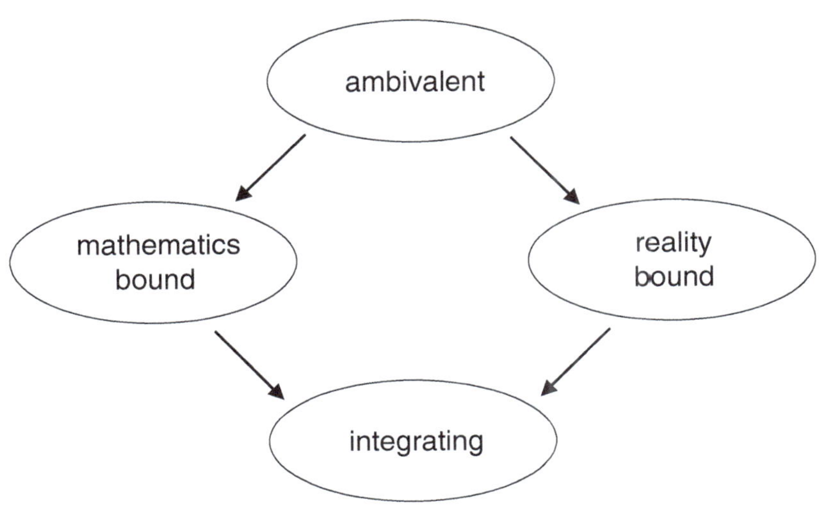

Fig. 12 Different kinds of dealing with context (Busse 2005, p. 356)

beliefs influence the modelling process enormously. However, these considerations should not hinder teachers in implementing modelling in their mathematics lessons, because many studies in the last decades have indicated that working on modelling problems leads to an increase in modelling competences (for example, Kaiser-Meßmer 1986; Kreckler 2015). Independent work on modelling problems and stimulating students' own activities are important for fostering students modelling competence. The supporting role of teachers is part of the next chapter.

2.10.2 The Role of Teachers in Modelling Processes

Implementing modelling into mathematics involves teachers as one of the focal points. They do not only have to be convinced of the usefulness of mathematical modelling; rather, they have to overcome suspected obstacles. Furthermore, their attitude can influence their way of supporting students and their decision as to which detail of the modelling process they select as the subject of discussion. Moreover, they have to know how to support students' working process best. In order to do so, special competencies are necessary. In the following, research results concerning the role of teachers in and their influence on modelling processes are presented.

Suspected obstacles are one reason for teachers not to implement mathematical modelling in their lessons. Blum (1996) differentiated these obstacles into four categories: organizational obstacles (especially shortage of time), student-related obstacles (modelling is assumed to be too difficult for students), teacher-related obstacles (not enough time for adapting tasks and preparing them in detail, lack of required skills), and material-related obstacles (knowledge of only a few modelling problems suitable for their lessons). However, these categories did not come out of empirical analysis. In 2008, Schmidt (2011) conducted a study with 101 teachers from primary and secondary school to find out whether the obstacles Blum categorised could be identified empirically (or had changed during time). The teachers named three main obstacles: lack of time, complexity of performance assessment, and lack of material. The first obstacle, lack of time, could be differentiated into lack of time necessary for working on modelling problems in the classroom and lack of time for preparation of modelling lessons. Teachers often expressed a desire not to waste time by working on modelling problems, but needed to fulfil the curriculum. This is astonishing, because modelling has been part of the curriculum in every German state for nearly a decade then (see Sect. 2.7). Concerning the last obstacle, lack of material, modelling problems for students in Grades 8–13 especially were mentioned. Whereas in the above study this obstacle could be overcome by presenting modelling problems to the teachers, the other two obstacles seemed to be more resistant. The teacher training that took place within the framework of the study did not change teachers' attitudes towards the other two obstacles: Even after the teacher training, teachers still found it difficult to assess modelling problems.

As stated above, students' beliefs and thinking styles can influence their modelling process. Similar findings about *teachers' beliefs and thinking styles* were identified: Teachers emphasised different features of the modelling process in reference to their mathematical thinking style or preferred way of representation. By analysing videotaped lessons of three different teachers, Borromeo Ferri (2011) found three different types of teachers. Some of the teachers underlined formal aspects while supporting students during their modelling process and discussions about solutions of modelling problems, whereas others emphasised reality-related aspects in order to validate the results and help students. A third type considered both formal mathematical aspects as well as real-world aspects. It is important to note that teachers are often not conscious of their own behaviour concerning this aspect. However, they certainly influenced the students' handling of modelling problems (Borromeo Ferri 2011; Borromeo Ferri and Blum 2013).

Not only teachers' priorities concerning modelling but also their behaviour in classes has an effect on students' modelling performance. Their *interventions* can hinder as well as support students' independent work on modelling problems. For independent work on modelling problems it is crucial to guide students as much as necessary and as little as possible (principle of minimal help, Aebli 1997). A well-known distinction between different kinds of interventions is the Zech's (2002) taxonomy of assistance. This method differentiates motivational, feedback, general-strategic, content-oriented strategic, and content-oriented assistance. The intensity of the intervention increases gradually from motivational assistance to content-oriented assistance. For complex problems such as modelling problems, the answer to the question of whether an intervention is appropriate or not is not that easy. Based on Zech's categorisation, Leiß created a descriptive analysis of adaptive teacher intervention in the modelling process. Here the analysed interventions were classified by trigger, level, and intention (see Leiß 2007).

Among others, the main results of Leiß's study illustrated that strategic interventions were included in the intervention repertoire of the observed teachers only very marginally and that teachers often chose indirect advice in situations where students had to find only one step by themselves in order to overcome the difficulty. Furthermore, only very few could be classified as adaptive and diagnosis based (Tropper et al. 2015). However, further studies (such as Link 2011; Stender and Kaiser 2015) did not confirm these results. In contrast, these studies provided evidence that specifically strategic interventions also have the potential of being adaptive and leading to metacognitive activities (see Link 2011).

Nevertheless, there is very little empirical knowledge about the effectiveness of single interventions. Stender (2016) investigated which kinds of scaffolding and intervention activities are adequate to promote independent students' modelling activities. In the framework of modelling days in Hamburg (see Sect. 2.9), the interventions of 10 future teachers supporting 45 students were analysed. Students worked on a complex, realistic, authentic modelling problem over three days. The pre-service teachers were trained to support the students merely by strategic interventions beforehand. The whole working processes were videotaped. On the basis of the analysis of 238 interventions, Stender and Kaiser emphasised the potential of interventions that are introduced ad hoc and asked the students to explain the state of work: On the one hand, students' answers gave possibilities for the teachers to diagnose possible difficulties. On the other hand, the students themselves structured their work while explaining and sometimes overcoming the difficulty without further help from the teachers (Stender and Kaiser 2015).

A different kind of support is *feedback*. The influence of different kinds of feedback on students' achievement and motivational variables was investigated in the framework of the Co^{2}Ca Project (Besser et al. 2015). The aim of this study was to determine a way for student performance to be assessed and reported that would enable teachers to analyse students' outcomes appropriately. The instrument for giving feedback needed to be both manageable for teachers and understandable for students. The investigation phase was divided into several parts: First, items were developed for the specific content areas and their related competencies. In addition, during piloting the tasks, types of feedback were first empirically tested and analysed. Second, a laboratory experiment followed in which different types of skill-based feedback on student performance were tested. In a third step, the experiences of the laboratory study were used in an empirical field study. Finally, a transfer study was carried out in which the influence of teacher training on the development of teachers' assessment competency was investigated. These studies showed that verbal feedback combined with various teacher- and mark-centred forms of assessment dominated as the most common forms of teacher feedback. Forms of self- or peer-evaluation were rare, but they were comparatively common among teachers who were well acquainted with diagnostic questions. In multi-level models, relationships between motivation and performance of students were identified: teacher- and mark-centred assessment practices were accompanied by lower motivation, whereas an ipsative reference standard orientation of the teacher was accompanied by increased motivation. Thus, the teachers' diagnostic skills were connected with better test scores of students. As expected, different types of feedback (process-related feedback, social-comparative feedback, and criteria-based feedback were used in the study) resulted in different effects on student motivation and on the attribution of test results. The criteria-based feedback had comparatively positive effects. Overall, on a quantitative level no significant improvements in performance were identified. Furthermore, first results of the teacher training study indicate that teachers who took part in the teacher training outperformed those who had not been trained in formative assessment (Klieme et al. 2010; Besser et al. 2015).

Mathematical modelling is not compulsory content in teacher education programmes at universities in German-speaking countries. Only at some universities (e.g., Hamburg and Kassel) are courses offered regularly. Often, these courses are linked to practices such as the above mentioned modelling days (see Sect. 2.9). If future teachers need be enabled to implement mathematical modelling in their future teaching, the conceptions of such seminars have to be based on considerations about necessary teacher competencies for modelling. Borromeo Ferri and Blum (2010) distinguish between five different categories of *teacher competencies for modelling*:

(1) Theory-oriented competency (contains necessary knowledge about theoretical aspects of modelling such as knowledge about modelling cycles, goals and perspectives for modelling, types of modelling tasks, and theoretical considerations about modelling competencies).
(2) Task-related competency (contains ability to solve a modelling problem, to analyse possible barriers and necessary competencies, and to create modelling tasks on their own).
(3) Teaching competency (contains micro- and macro-scaffolding abilities such as the ability to plan and perform modelling lessons and knowledge of appropriate adaptive interventions to enable students to work as independently as possible)
(4) Diagnostic competency (contains the ability to identify phases in students' modelling processes and to diagnose students' difficulties during such processes in order to support students during their work and to select modelling problems).
(5) Assessment competency (contains the ability to construct appropriate tasks and tests for assessing students' modelling competencies as well as assessing students' work on modelling problems).

The fifth dimension is not considered to be reasonable for teacher education at university due to time restrictions and students' experience. An example of such seminars as well as the evaluation can be found in Borromeo Ferri and Blum (2010).

Due to the fact that mathematics teachers often do not know how to implement mathematical modelling in their classroom and often assume that there are obstacles as mentioned above, courses for practicing teachers are necessary. One example of such a course is the teacher training course developed in the framework of the international project LEMA (Learning and Education in and through Modelling and Applications). On the basis of a requirement analyses as well as on theoretical considerations, five key modules were developed, implemented, and evaluated. The evaluation shows that the course had strong positive effects on the teachers' pedagogical content knowledge and self-efficacy in terms of modelling, but no positive effects on the teachers' biases (Maaß and Gurlitt 2011).

As shown above, teachers have a great influence on students' modelling processes, although they are often unaware of their impact. It has also become clear that many competencies are necessary in order to support students as appropriately as possible and in order to implement modelling activities adequate for mathematics lessons. In the discussion of scaffolding, these interactions that can foster or hinder students' independent work on modelling problems are part of micro-scaffolding. All aspects that can be arranged and planned before are called macro-scaffolds (Hammond and Gibbons 2005). Results concerning aspects of macro-scaffolding are presented in the next chapter.

2.10.3 Classroom Settings

As shown above, teachers play an important role in implementing mathematical modelling successfully into mathematic lessons and in fostering students' modelling competencies. Furthermore, classroom settings (which can surely be established by teachers as well) play an important role. So apart from direct teacher behaviour, the design of single modelling lessons as well as the whole modelling teaching unit (both of which are typically arranged by teachers) have been in the focus of research as well.

In the DISUM project, a *directive teaching approach* (i.e., *teacher-centred*) was contrasted with an *operative-strategic teaching approach* (i.e., more *student-centred*) during a 10-lesson learning unit. The study was carried out in 18 classes of Grade 9. The results clearly indicate the advantages of operative-strategic teaching in terms of the increase in students' modelling competence as well as their self-regulation (Schukajlow et al. 2012). However, working completely independently in groups on modelling problems—the third evaluated teaching approach—did not allow students to tackle the modelling problem successfully (Schukajlow and Messner 2007). This outcome underlines the important role of teachers in fostering students' modelling competence and the necessity of directive phases in operative-strategic teaching.

In the framework of the same project, the *influence of class sizes* that were taught in an operative-strategic way was investigated as well. Seven classes were of "normal" size for German standards ($\sim$26 students per class) and five were "small" classes ($\sim$16 students per class). The results show that modelling competence can be fostered in smaller classes significantly better than in classes of standard German size smaller ones, but in both classes, student modelling competences increased during the 10-lesson teaching unit (Schukajlow and Blum 2011).

Again in the DISUM framework, a third factor was tested that may influence the students' work on modelling problems and give them support in solving modelling problems independently. During a two-day intervention in six classes of Grade 9, a *solution plan* was introduced as a scaffold (Blum 2011). This plan was comprised of four stages: understanding the task, establishing the model, using mathematics, and explaining the results. Each stage was explained to students with two explicative bullet points. This plan was a variation of the four-step modelling cycle and included some hints about what to do in the different steps. It was not meant as schema for solving modelling problems but as an aid. The results show the potential of the solution plan as guideline: The students using the solution plan while working on the modelling problem reported that they used strategies more frequently than those of the control group. Furthermore, students using the solution plan showed higher achievement than those in the other group (Schukajlow et al. 2010, 2015a, b).

Supporting students' modelling processes most effectively can be a great challenge for teachers. In order to have enough time to support students individually, measures of support that can be prepared beforehand are of high interest.

2.10.4 Design of Modelling Problems

The design of a modelling problem plays an important role in the modelling process and can influence students' work on the problem. As shown in the example above, the context of a modelling problem has a great influence on the students' working process.

The results of research into the design of modelling problems can be distinguished between results concerning the characteristics (and thus impact) of single problems and results concerning those of a set of problems. Furthermore, one can distinguish between impact on students' working behaviour and students' modelling competence.

Kaiser's (1995) study on modelling problems in general described theoretically different potential impacts that modelling problems could have. One potential impact is the possibility of developing a *personal meaning* for mathematics. In an empirical study with 15 students in Grade 10, Vorhölter analysed the role of modelling problems in constructing a personal meaning for mathematics. In general, 12 different personal meanings emerged from the interviews, which were grouped into five areas: (1) as a tool for life, (2) for getting social appreciation, (3) for getting satisfaction, (4) consideration about mathematics lessons, and (5) concerning mathematical knowledge. The most important personal meanings for the students were those of mathematics as a tool for life and for satisfaction. Often, however, it was not possible for the students to realise those personal meanings, i.e., they were not told and were not able to determine for themselves how they could use the mathematics they had learnt as a tool. Lessons involving modelling, however, helped students realise these two important personal meanings more often. It was not only the context of the modelling tasks that helped the students to realise their personal meaning, however; other characteristics of modelling tasks (such as openness and the challenge to develop one's own approach) as well the setting (for example, group work or different teacher behaviour) helped the students achieve their own personal meaning (Vorhölter 2009).

Kaiser (1995) also showed that modelling problems also have the potential to *motivate* students. This hypothesis was reassessed in the STRATUM Project. Within the projects' framework, 13 teaching units were developed for underachieving students. The 959 participating students and 54 participating teachers were divided into two intervention groups and one control group. In terms of various variables, students' motivation was measured before and after the teaching unit. The results of the study partly confirmed Kaiser's hypothesis: Students' motivation did not increase, but the decrease of learning motivation could be blocked in the intervention groups (Maaß and Mischo 2012). Kreckler (2015) confirmed this result in a certain way: The majority of the 332 participating students of her study wished to work on modelling problems during mathematics lessons more often, irrespective of gender, mathematical competence, and mathematical theme. Moreover, the four-lesson teaching unit in the framework of Kreckler's project resulted in a sustainable increase in modelling competence.

As indicated above, in the last years several studies have been carried out with the intention of determining how to optimally promote students' modelling competencies. The projects focused on different groups of students as well as different activities. In all these studies, sets of modelling problems were developed.

One of the teaching approaches developed especially for novice modellers is the computer-based learning environment KOMMA. The learning environment comprises four *heuristic worked-out examples*. In these examples, two fictional characters solved a modelling problem and explained their ideas, heuristic strategies, and heuristic tools. All the examples being worked out were structured using a 3-step modelling cycle. The modelling competence of the 316 participating eighth grade students were tested before, just after, and four months after the intervention. The results indicated a significant increase in modelling competence just after the implementation of the learning environment and lesser long-term effects. Underachieving students in particular benefited from the approach (Zöttl et al. 2010).

In another study, the examples being worked out were used as scaffolds. The interactions of four ninth grade students and their imitation of demonstrated behaviour in the examples were examined. The study points out that the number of imitations per student was quite different and that some elements were not imitated at all. Altogether, the examples' potential for helping students to work on modelling problems on their own became obvious. In contrast to the potential support of a teacher, examples can only provide solutions at a strategic level (Tropper et al. 2015).

In addition to the KOMMA Project, the ERMO Project focuses on novice student modellers and the fostering of their modelling competence as target. The effectiveness of two different approaches (a *holistic* as well as an *atomistic* approach; see Blomhøj and Jensen 2003) was tested against each other in the following way: The participating 15 ninth grade classes were divided into two groups. Each group was assigned five modelling problems that had the same context, but students' work on the problems differed: Whereas the students of the atomistic group only had to work on one step of the modelling cycle, the students of the holistic group had to go through the whole modelling process for every problem. The students' modelling competence was tested before and after the intervention unit as well as a half year after. The results indicated the strengths and weaknesses of both approaches, whereas both approaches are reasonably effective at fostering students' modelling competencies. However, the holistic approach was proven to be more effective for students with weaker performance in mathematics (Kaiser and Brand 2015).

In the framework of the MultiMa Project, the influence of demanding *multiple solutions* for one modelling problem was tested. Two groups of 144 ninth graders in six classes were compared. One of the groups was asked to work on a problem without having to make assumptions in order to solve the problem. In the other group, different assumptions were requested and students had to develop at least two different ones. Before and after the teaching unit, students were asked to self-report on their planning and monitoring strategies. The results of this study showed a positive influence on students' planning and monitoring in the group that were asked to develop multiple solutions (Schukajlow and Krug 2013).

Furthermore, prompting students to develop multiple solutions had no direct influence on their direct performance, but increased the number of developed solutions (Schukajlow et al. 2015a, b).

Overall, an appropriate complexity of tasks increasing within a set of modelling tasks is recommended (Maaß 2006; Blum 2011). Furthermore, a broad variation of contexts as well as mathematical domains is needed in order to guide students to transfer modelling strategies from one task to another (Blum 2011, 2015).

3 Summary and Looking Ahead

As presented above, modelling and applications were and still are an important part of German research on mathematics education. In the last century, the German discussion on modelling focused on conceptual aspects and exemplarily modelling problems. This was an important step in clarifying the content of the concept *mathematical model*. During this time, a discussion on different types of models and modelling examples in the light of a long German tradition of applications in school mathematics took place. An important step in bringing research and school practice closer together and integrating modelling examples into the classroom was the establishment of the German-speaking ISTRON group 25 years ago. A new development in integrating applications and modelling in all types of schools started in the last decades of the 20th century. A much-debated question is the adaptation of a particular modelling cycle for a particular research question. This development led to a greater internationalisation of German research on modelling and integration of modelling as a competency into the curriculum at the beginning of this millennium. Nowadays, modelling is part of the German national curriculum. However, as in most countries, applications and modelling play only a small role in everyday teaching. The presented empirical results show the main foci of the research on modelling in application in the last years. Currently, the effective promotion of students' modelling competencies is the core of research. Concurrently, instruments for helping students to work on modelling problems independently (and relieving teachers in some way) are being developed and analysed.

- The long tradition of applications in school mathematics in German-speaking countries is discussed.
- Approaches for the integration of modelling problems in school practice are described.
- The integration of modelling as a competency in the current educational standards is described.
- The influence of digital tools on school practice and research projects on mathematical modelling is described.
- New empirical research projects on mathematical modelling in German-speaking countries on the role of students and teachers, classroom settings, and design of modelling problems are put forward.

Open Access This chapter is distributed under the terms of the Creative Commons Attribution 4.0 International License (http://creativecommons.org/licenses/by/4.0/), which permits use, duplication, adaptation, distribution and reproduction in any medium or format, as long as you give appropriate credit to the original author(s) and the source, provide a link to the Creative Commons license and indicate if changes were made.

The images or other third party material in this chapter are included in the work's Creative Commons license, unless indicated otherwise in the credit line; if such material is not included in the work's Creative Commons license and the respective action is not permitted by statutory regulation, users will need to obtain permission from the license holder to duplicate, adapt or reproduce the material.

References

Aebli, H. (1997). *Zwölf Grundformen des Lehrens: Eine allgemeine Didaktik auf psychologischer Grundlage* (9th ed.). Stuttgart: Klett-Cotta.

Ahrens, W. (1904). *Scherz und Ernst in der Mathematik*. Hildesheim: Olms-Weidmann.

Bardy, P., Danckwerts, R., & Schornstein, J. (Eds.) (1996). *Materialien für einen realitätsbezogenen Mathematikunterricht, Band 3*. Hildesheim: Franzbecker.

Barzel, B., Hußmann, S., & Leuders, T. (2005). *Computer, Internet & Co im Mathematikunterricht*. Berlin: Cornelsen Scriptor.

Besser, M., Blum, W., & Leiß, D. (2015). How to support teachers to give feedback to modelling tasks effectively? results from a teacher-training-study in the Co2CA project. In G. A. Stillman, W. Blum, & M. Salett Biembengut (Eds.), *Mathematical modelling in education research and practice* (pp. 151–160). Cham: Springer International Publishing.

Blomhøj, M., & Jensen, T. H. (2003). Developing mathematical modelling competence: conceptual clarification and educational planning. *Teaching Mathematics and its Applications, 22*(3), 123–139.

Blum, W. (1978). Einkommensteuern als Thema des Analysisunterrichts in der beruflichen Oberstufe. *Die berufsbildende Schule, 30*(11), 642–651.

Blum, W. (1985). Anwendungsorientierter Mathematikunterricht in der didaktischen Diskussion. *Mathematische Semesterberichte, 32*(2), 195–232.

Blum, W. (Ed.). (1993). *Anwendungen und Modellbildung im Mathematikunterricht*. Hildesheim: Franzbecker.

Blum, W. (1996). Anwendungsbezüge im Mathematikunterricht—Trends und Perspektiven. In G. Kadunz, H. Kautschitsch, G. Ossimitz, & E. Schneider (Eds.), *Trends und Perspektiven* (pp. 15–38). Wien: Hölder-Pichler-Tempsky.

Blum, W. (2011). Can modelling be taught and learnt? some answers from empirical research. In G. Kaiser, W. Blum, R. Borromeo Ferri, & G. A. Stillman (Eds.). *Trends in teaching and learning of mathematical modelling. ICTMA14* (pp. 15–30). Dordrecht: Springer.

Blum, W. (2015). Quality teaching of mathematical modelling: What do we know, what can we do? In S. J. Cho (Ed.), *The Proceedings of the 12th International Congress on Mathematical Education* (pp. 73–96). Cham: Springer International Publishing.

Blum, W., Berry, J., Biehler, R., Huntley, I., Kaiser-Meßmer, G., & Profke, L. (Eds.) (1989). Applications and modelling in learning and teaching mathematics. Chichester: Ellis Horwood.

Blum, W., Galbraith, P. L., Henn, H.-W., & Niss, M. (Eds.). (2007). *Modelling and applications in mathematics education. The 14th ICMI study*. New York: Springer.

Blum, W., & Kaiser, G. (1984). Analysis of applications and of conceptions for an application-oriented mathematics instruction. In J. S. Berry, D. Burghes, I. Huntley, D. James, & A. Moscardini (Eds.), *Teaching and applying mathematical modelling* (pp. 201–214). Chichester: Horwood.

Blum, W., & Kirsch, A. (1989). The problem of the graphic artist. In W. Blum, J. S. Berry, R. Biehler, I. D. Huntley, G. Kaiser-Meßmer, & L. Profke (Eds.), *Applications and modelling in learning and teaching mathematics* (pp. 129–135). Chichester: Ellis Horwood.

Blum, W., & Leiß, D. (2005). Modellieren im Unterricht mit der "Tanken"-Aufgabe. *mathematik lehren, 128*, 18–21.

Blum, W., & Leiß, D. (2007). How do students and teachers deal with mathematical modelling problems? The example sugarloaf and the DISUM project. In C. Haines, P. L. Galbraith, W. Blum, & S. Khan (Eds.), *Mathematical modelling (ICTMA 12). Education, engineering and economics* (pp. 222–231). Chichester: Horwood.

Blum, W., & Niss, M. (1991). Applied mathematical problem solving, modelling, applications and links to other subjects—State, trends and issues in mathematics instruction. *Educational Studies in Mathematics, 22*(1), 37–68.

Blum, W., & Törner, G. (1983). *Didaktik der Analysis*. Göttingen: Vandenhoeck & Ruprecht.

Böer, H. (1993). Extremwertproblem Milchtüte. Eine tatsächliche Problemstellung aktueller industrieller Massenproduktion. In W. Blum (Ed.), *Anwendungen und Modellbildung im Mathematikunterricht* (pp. 1–16). Hildesheim: Franzbecker.

Böhm, U. (2013). *Modellierungskompetenzen langfristig und kumulativ fördern: Tätigkeitstheoretische Analyse des mathematischen Modellierens in der Sekundarstufe I.* Wiesbaden: Springer Spektrum.

Borromeo Ferri, R. (2004). Vom Realmodell zum mathematischen Modell - Analyse von Übersetzungsprozessen aus der Perspektive mathematischer Denkstile. *Beiträge zum Mathematikunterricht*, 109–112.

Borromeo Ferri, R. (2006). Theoretical and empirical differentiations of phases in the modelling process. *ZDM—The International Journal on Mathematics Education, 38*(2), 86–95.

Borromeo Ferri, R. (2010). On the influence of mathematical thinking styles on learners' modeling behavior. *Journal für Mathematik-Didaktik, 31*(1), 99–118.

Borromeo Ferri, R. (2011). *Wege zur Innenwelt des mathematischen Modellierens: Kognitive Analysen zu Modellierungsprozessen im Mathematikunterricht*. Wiesbaden: Vieweg + Teubner Verlag/Springer Fachmedien.

Borromeo Ferri, R., & Blum, W. (2010). Mathematical Modelling in teacher education - experiences from a modelling seminar. In V. Durand-Guerrier, S. Soury-Lavergne, & F. Arzarello (Eds.), *CERME 6. Proceedings of the sixth congress of the European Society for Research in Mathematics Education, January 28th-February 1st 2009 Lyon (France)* (pp. 2046–2055). Lyon: Institut national de recherche pédagogique.

Borromeo Ferri, R., & Blum, W. (2013). Insights into teachers' unconscious behaviour in modeling contexts. In R. Lesh, P. L. Galbraith, C. R. Haines, & A. Hurford (Eds.), *Modeling students' mathematical modelling competencies. ICTMA 13* (pp. 423–432). New York, NY: Springer.

Breidenbach, W. (1969). *Methodik des Mathematikunterrichts in Grund- und Hauptschulen*. Hannover: Schroedel.

Büchter, A., & Leuders, T. (2005). *Mathematikaufgaben selbst entwickeln. Lernen fördern Leistung überprüfen*. Berlin: Cornelsen Scriptor.

Burscheid, H. (1980). Beiträge zur Anwendung der Mathematik im Unterricht. Versuch einer Zusammenfassung. *Zentralblatt für Didaktik der Mathematik, 12*, 63–69.

Burton, L. (2004). What does it mean to be a mathematical enquirer—Learning as research. In L. Burton (Ed.), *Mathematicians as enquirers. Learning about learning mathematics* (pp. 177–204). Boston: Kluwer Academic Publishers.

Busse, A. (2005). Individual ways of dealing with the context of realistic tasks—first steps towards a typology. *ZDM, 37*(5), 354–360.

Ebenhöh, W. (1990). Mathematische Modellierung Grundgedanken und Beispiele. *Der Mathematikunterricht, 36*(4), 5–15.

Fischer, R., & Malle, G. (1985). *Mensch und Mathematik*. Mannheim: Bibliographisches Institut.

Franke, M., & Ruwisch, S. (2010). *Didaktik des Sachrechnens in der Grundschule*. Heidelberg: Spektrum.

Freudenthal, H. (1978). *Vorrede zu einer Wissenschaft vom Mathematikunterricht*. Oldenbourg.

Kaiser-Meßmer, G., Blum, W., & Schober, M. (1992). *Dokumentation ausgewählter Literatur zum anwendungsorientierten Mathematikunterricht. Teil 2, 1982–1989*. Karlsruhe: Fachinformationszentrum Karlsruhe.

Geiger, V. (2011). Factors affecting teachers' adoption of innovative practices with technology and mathematical modelling. In G. Kaiser, W. Blum, R. Borromeo Ferri & G. Stillman (Eds.), *Trends in teaching and learning of mathematical modelling* (pp. 305–314). Dordrecht: Springer.

Gellert, U., Jablonka, E., & Keitel, C. (2001). Mathematical literacy and common sense in mathematics education. In B. Atweh, H. Forgasz, & B. Nebres (Eds.), *Sociocultural research on mathematics education* (pp. 57–76). Mahwah: Erlbaum.

Greefrath, G. (2004). Offene Aufgaben mit Realitätsbezug. Eine Übersicht mit Beispielen und erste Ergebnisse aus Fallstudien. *Mathematica Didactica, 2*(27), 16–38.

Greefrath, G. (2010). *Didaktik des Sachrechnens in der Sekundarstufe*. Heidelberg: Springer Spektrum.

Greefrath, G. (2011). Using technologies: New possibilities of teaching and learning modelling—Overview. In G. Kaiser, W. Blum, R. Borromeo Ferri, G. Stillman (Eds.), *Trends in teaching and learning of mathematical modelling, ICTMA 14* (pp. 301–304), Dordrecht: Springer.

Greefrath, G. (2015). Problem solving methods for mathematical modeling. In G. Stillman, W. Blum, M. S. Biembengut (Eds.), *Mathematical modelling in education research and practice. Cultural, social and cognitive influences ICTMA 16* (pp. 173–183). Dordrecht, Heidelberg, London, New York: Springer.

Greefrath, G., Kaiser, G., Blum, W., & Borromeo Ferri, R. (2013). Mathematisches Modellieren—Eine Einführung in theoretische und didaktische Hintergründe. In R. Borromeo Ferri, G. Greefrath, G. Kaiser (Eds.). *Mathematisches Modellieren für Schule und Hochschule. Theoretische und didaktische Hintergründe*. Wiesbaden: Springer Spektrum.

Greefrath, G., & Mühlenfeld, U. (2007). *Realitätsbezogene Aufgaben für die Sekundarstufe II*. Troisdorf: Bildungsverlag Eins.

Greefrath, G., Siller, H.-S., & Blum, W. (2016). 25 Jahre ISTRON – 25 Jahre Arbeit für einen realitätsbezogenen Mathematikunterricht. *Mitteilungen der Gesellschaft für Didaktik der Mathematik, 100*, 19–22.

Greefrath, G., Siller, H.-J., & Weitendorf, J. (2011). Modelling considering the influences of technology. In G. Kaiser, W. Blum, R. Borromeo Ferri & G. Stillman (Eds.), *Trends in teaching and learning of mathematical modelling* (pp. 315–330). Dordrecht: Springer.

Greefrath, G., & Weigand, H.-G. (2012). Simulieren—mit Modellen experimentieren. *mathematik lehren, 174*, 2–6.

Griesel, H. (2005). Modelle und Modellieren eine didaktisch orientierte Sachanalyse, zugleich ein Beitrag zu den Grundlagen einer mathematischen Beschreibung der Welt. In H.-W. Henn, & G. Kaiser (Eds.), *Mathematikunterricht im Spannungsfeld von Evolution und Evaluation. Festschrift für Werner Blum* (pp. 61–70). Hildesheim: div.

Hammond, J., & Gibbons, P. (2005). Putting scaffolding to work: The contribution of scaffolding in articulating ESL education. *Prospect, 20*(1), 6–30.

Hartmann, B. (1913). *Der Rechenunterricht in der deutschen Volksschule vom Standpunkte des erziehenden Unterrichts*. Leipzig: Kesselring.

Henn, H.-W. (1980). Die Theorie des Regenbogens als Beispiel für beziehungshaltige Analysis im Oberstufenunterricht. *Journal für Mathematikdidaktik, 1*, 62–85.

Henn, H.-W. (1995). Volumenbestimmung bei einem Rundfass. In G. Graumann, T. Jahnke, G. Kaiser, & J. Meyer (Eds.), *Materialien für einen realitätsbezogenen Mathematikunterricht Bd. 2 (ISTRON)* (pp. 56–65). Hildesheim: Franzbecker.

Henn, H.-W. (1998). The Impact of Computer Algebra Systems on Modelling Activities. In P. Galbraith, W. Blum, G. Book & I.D. Huntley (Eds.), *Mathematical modelling. Teaching and assessment in a technology-rich world* (pp. 115–124). Chichester: Horwood.

Henn, H.-W. (2002). Mathematik und der Rest der Welt. *mathematik lehren, 113*, 4–7.

Henn, H.-W. (2007). Modelling pedagogy—Overview. In W. Blum, P.L. Galbraith, H,-W. Henn & M. Niss (Eds.). *Modelling and applications in mathematics education. The 14th ICMI Study* (pp. 321–324). New York: Springer.

Henn, H.-W., & Maaß, K. (2003). Standardthemen im realitätsbezogenen Mathematikunterricht. In W. Henn & K. Maaß (Eds.), *Materialien für einen realitätsbezogenen Mathematikunterricht. Bd. 8 (ISTRON)* (pp. 1–5). Hildesheim: Franzbecker.

Herget, W., & Scholz, D. (1998). *Die etwas andere Aufgabe aus der Zeitung*. Seelze: Kallmeyer.

Hertz, H. (1894). *Die Prinzipien der Mechanik in neuem Zusammenhange dargestellt*. Leipzig: Barth.

Hischer, H. (2002). *Mathematikunterricht und Neue Medien. Hintergründe und Begründungen in fachdidaktischer und fachübergreifender Sicht*. Hildesheim: Franzbecker.

Humenberger, H., & Reichel, H.-C. (1995). *Fundamentale Ideen der Angewandten Mathematik*. Mannheim, Leipzig, Wien, Zürich: BI Wissenschaftsverlag.

Kaiser, G. (1995). Realitätsbezüge im Mathematikunterricht: Ein Überblick über die aktuelle und historische Diskussion. In G. Graumann, T. Jahnke, G. Kaiser, & J. Meyer (Eds.), *Schriftenreihe der Istron-Gruppe: Materialien für einen realitätsbezogenen Mathematikunterricht* (Vol. 2, pp. 66–81). Hildesheim: Franzbecker.

Kaiser, G. (2005). Mathematical modelling in school—Examples and experiences. In Henn, H.-W.; Kaiser, G. (Eds.), *Mathematikunterricht im Spannungsfeld von Evolution und Evaluation. Festband für Werner Blum*. Hildesheim: Franzbecker.

Kaiser, G. (2007). Modelling and modelling competencies in school. In C. Haines, P. Galbraith, W. Blum & S. Khan (Eds.), *Mathematical modelling (ICTMA 12). Education, engineering and economics* (pp. 110–119). Chichester: Horwood.

Kaiser, G. (2015). Werner Blum und sein Beitrag zum Lehren und Lernen mathematischen Modellierens. In G. Kaiser & H.-W. Henn (Eds.), *Werner Blum und seine Beiträge zum Modellieren im Mathematikunterricht* (pp. 1–24). Wiesbaden: Springer Spektrum.

Kaiser, G., Blum, W., Borromeo Ferri, R., & Greefrath, G. (2015). Anwendungen und Modellieren. In R. Bruder, L. Hefendehl-Hebeker, B. Schmidt-Thieme, & H.-G. Weigand (Eds.), *Handbuch der Mathematikdidaktik* (pp. 357–383). Berlin: Springer.

Kaiser, G., Blum, W., & Schober, M. (unter Mitarbeit von Stein, R.) (1982). *Dokumentation ausgewählter Literatur zum anwendungsorientierten Mathematikunterricht*. Karlsruhe: Fachinformationszentrum Energie, Physik, Mathematik.

Kaiser, G., Bracke, M., Göttlich, S., & Kaland, C. (2013). Authentic complex modelling problems in mathematics education. In A. Damlamian, J. F. Rodrigues, & R. Sträßer (Eds.), *New ICMI Study Series. Educational interfaces between mathematics and industry* (Vol. 16, pp. 287–297). Cham: Springer.

Kaiser, G., & Brand, S. (2015). Modelling competencies: Past development and further perspectives. In G. A. Stillman, W. Blum, & M. Salett Biembengut (Eds.), *Mathematical modelling in education research and practice* (pp. 129–149). Cham: Springer International Publishing.

Kaiser, G., & Schwarz, B. (2010). Authentic modelling problems in mathematics education—Examples and experiences. *Journal für Mathematik-Didaktik, 31*(1), 51–76.

Kaiser, G., & Sriraman, B. (2006). A global survey of international perspectives on modelling in mathematics education. *ZDM, 38*(3), 302–310.

Kaiser, G., & Stender, P. (2013). Complex modelling problems in co-operative, self-directed learning environments. In G. Stillman, G. Kaiser, W. Blum, & J. Brown (Eds.), *Teaching mathematical modelling: Connecting to research and practice* (pp. 277–293). Dordrecht: Springer.

Kaiser-Meßmer, G. (1986). *Anwendungen im Mathematikunterricht. Vol. 1 & 2*. Bad Salzdetfurth: Franzbecker.

Klein, F. (1907). *Vorträge über den mathematischen Unterricht an den höheren Schulen. Teil 1*. Leipzig: Teubner.

Klieme, E., Bürgermeister, A., Harks, B., Blum, W., Leiß, D., & Rakoczy, K. (2010). Leistungsbeurteilung und Kompetenzmodellierung im Mathematikunterricht. In E. Klieme (Ed.), *Zeitschrift für Pädagogik Beiheft: Vol. 56. Kompetenzmodellierung. Zwischenbilanz des DFG-Schwerpunktprogramms und Perspektiven des Forschungsansatzes.* Weinheim: Beltz.

KMK. (2012). Bildungsstandards im Fach Mathematik für die Allgemeine Hochschulreife. *Beschluss der Kultusministerkonferenz vom, 18*(10), 2012.

Kreckler, J. (2015). *Standortplanung und Geometrie*. Wiesbaden: Springer Fachmedien Wiesbaden.

Kühnel, J. (1916). *Neubau des Rechenunterrichts*. Leipzig: Klinkhardt.

Leiß, D. (2007). *Hilf mir, es selbst zu tun: Lehrerinterventionen beim mathematischen Modellieren*. Hildesheim: Franzbecker.

Leiß, D., Schukajlow, S., Blum, W., Messner, R., & Pekrun, R. (2010). The role of the situation model in mathematical modelling—Task analyses, student competencies, and teacher interventions. *Journal für Mathematik-Didaktik, 31*(1), 119–141.

Lietzmann, W. (1919). *Methodik des mathematischen Unterrichts, I. Teil*. Leipzig: Quelle & Meyer.

Link, F. (2011). *Problemlöseprozesse selbstständigkeitsorientiert begleiten: Kontexte und Bedeutungen strategischer Lehrerinterventionen in der Sekundarstufe I.* Wiesbaden: Vieweg + Teubner Verlag/Springer Fachmedien Wiesbaden GmbH Wiesbaden.

Maaß, K. (2002). Handytarife. *Mathematik Lehren, 113*, 53–57.

Maaß, K. (2004). *Mathematisches Modellieren im Unterricht – Ergebnisse einer empirischen Studie*. Hildesheim: Franzbecker.

Maaß, K. (2005). Modellieren im Mathematikunterricht der Sekundarstufe I. *Journal für Mathematikdidaktik, 26*, 114–142.

Maaß, K. (2006). What are modelling competencies? *ZDM, 38*(2), 113–142.

Maaß, J. (2007). Ethik im Mathematikunterricht? Modellierung reflektieren! In G. Greefrath & J. Maaß (Eds.), *Materialien für einen realitätsbezogenen Mathematikunterricht. Vol. 11* (pp. 54–61). Hildesheim: Franzbecker.

Maaß, K. (2010). Classification scheme for modelling tasks. *Journal für Mathematik-Didaktik, 31*, 285–311.

Maaß, K., & Gurlitt, J. (2011). LEMA—Professional development of teachers in relation to mathematical modelling. In G. Kaiser, W. Blum, R. Borromeo Ferri, & G. A. Stillman (Eds.). *International Perspectives on the Teaching and Learning of Mathematical Modelling, Trends in teaching and learning of mathematical modelling. ICTMA14* (Vol. 1, pp. 629–639). Dordrecht: Springer.

Maaß, K., & Mischo, C. (2012). Fördert mathematisches Modellieren die Motivation in Mathematik? Befunde einer Interventionsstudie bei HauptschülerInnen. *Mathematica Didactica, 35*, 25–49.

Neunzert, H., & Rosenberger, B. (1991). *Schlüssel zu Mathematik*. Econ.

Niss, M. (2003). Mathematical competencies and the learning of mathematics: the Danish KOM project. In A. Gagatsis, & S. Papastavridis (Eds.), *Mediterranean Conference on Mathematical Education* (pp. 115–124). Athen: 3rd Hellenic Mathematical Society and Cyprus Mathematical Society.

Niss, M., Blum, W., & Galbraith, P. (2007). Introduction. In W. Blum, P.L. Galbraith, H.-W. Henn & M. Niss (Eds.), *Modelling and applications in mathematics education. The 14th ICMI Study* (pp. 3–32). New York: Springer.

Pollak, H. O. (1968). On some of the problems of teaching applications of mathematics. *Educational Studies in Mathematics, 1*(1/2), 24–30.

Pollak, H. O. (1977). The interaction between mathematics and other school subjects (including integrated courses). In H. Athen & H. Kunle (Eds.), *Proceedings of the Third International Congress on Mathematical Education* (pp. 255–264). Karlsruhe: Zentralblatt für Didaktik der Mathematik.

Savelsbergh, E. R., Drijvers, P. H. M., van de Giessen, C., Heck, A., Hooyman, K., Kruger, J., et al. (2008). *Modelleren en computer-modellen in de β-vakken: advies op verzoek van de gezamenlijke β-vernieuwingscommissies*. Utrecht: Freudenthal Instituut voor Didactiek van Wiskunde en Natuurwetenschappen.

Schmidt, B. (2011). Modelling in the classroom: Obstacles from the teacher's perspective. In G. Kaiser, W. Blum, R. Borromeo Ferri, & G. A. Stillman (Eds.). *International perspectives on the teaching and learning of mathematical modelling, trends in teaching and learning of mathematical modelling. ICTMA14* (Vol. 1, pp. 641–651). Dordrecht: Springer.

Schukajlow, S., & Blum, W. (2011). Zum Einfluss der Klassengröße auf Modellierungskompetenz, Selbst- und Unterrichtswahrnehmungen von Schülern in selbständigkeitsorientierten Lehr-Lernformen. *Journal für Mathematik-Didaktik, 32*(2), 133–151.

Schukajlow, S., Kolter, J., & Blum, W. (2015a). Scaffolding mathematical modelling with a solution plan. *ZDM, 47*(7), 1241–1254.

Schukajlow, S., Krug, A., & Rakoczy, K. (2015b). Effects of prompting multiple solutions for modelling problems on students' performance. *Educational Studies in Mathematics, 89*(3), 393–417.

Schukajlow, S., Krämer, J., Blum, W., Besser, M., Brode, R., Leiß, D., et al. (2010). Lösungsplan in Schülerhand: zusätzliche Hürde oder Schlüssel zum Erfolg? In A. Lindmeier & S. Ufer (Eds.), *Beiträge zum Mathematikunterricht*. WTM: Münster.

Schukajlow, S., & Krug, A. (2013). Planning, monitorin and multiple solutions while solving modelling problems. In A. M. Lindmeier & A. Heinze (Eds.), *Mathematics learning across the life span. Proceedings of the 37th Conference of the International Group for the Psychology of Mathematics Education; PME 37; Kiel, Germany, July 28–August 02, 2013* (pp. 177–184). Kiel: IPN Leibniz Institute for Science and Mathematics Education.

Schukajlow, S., & Leiß, D. (2011). Selbstberichtete Strategienutzung und mathematische Modellierungskompetenz. *Journal für Mathematikdidaktik, 32*, 53–77.

Schukajlow, S., Leiß, D., Pekrun, R., Blum, W., Müller, M., & Messner, R. (2012). Teaching methods for modelling problems and students' task-specific enjoyment, value, interest and self-efficacy expectations. *Educational Studies in Mathematics, 79*(2), 215–237.

Schukajlow, S., & Messner, R. (2007). Selbständiges Arbeiten mit Modellierungsaufgaben? Ja, aber wie?! In *Beiträge zum Mathematikunterricht* (pp. 370–373). Hildesheim: Franzbecker.

Schupp, H. (1988). Anwendungsorientierter Mathematikunterricht in der Sekundarstufe I zwischen Tradition und neuen Impulsen. *Der Mathematikunterricht, 34*(6), 5–16.

Schupp, H. (1989). Applied mathematics instruction in the lower secondary level—Between traditional and new approaches. In W. Blum, et al. (Eds.), *Applications and modelling in learning and teaching mathematics* (pp. 37–46). Chichester: Ellis Horwood.

Sonar, T. (2001). *Angewandte Mathematik, Modellbildung und Informatik*. Braunschweig: Vieweg.

Stender, P. (2016). *Wirkungsvolle Lehrerinterventionsformen bei komplexen Modellierungsaufgaben. Perspektiven der Mathematikdidaktik*. Wiesbaden: Springer Fachmedien.

Stender, P., & Kaiser, G. (2015). Scaffolding in complex modelling situations. *ZDM, 47*(7), 1255–1267.

Tropper, N., Leiß, D., & Hänze, M. (2015). Teachers' temporary support and worked-out examples as elements of scaffolding in mathematical modelling. *ZDM, 47*(7), 1225–1240.

Vorhölter, K. (2009). *Sinn im Mathematikunterricht* (Vol. 27). Opladen, Hamburg: Budrich.

Vorhölter, K., & Kaiser, G. (2016). Theoretical and pedagogical considerations in promoting students' metacognitive modeling competencies. In C. Hirsch (Ed.), *Annual perspectives in mathematics education 2016: Mathematical modeling and modeling mathematics* (pp. 273–280). Reston, VA: National Council of Teachers of Mathematics.

Vorhölter, K., Kaiser, G., & Borromeo Ferri, R. (2014). Modelling in mathematics classroom instruction: An innovative approach for transforming mathematics education. In Y. Li, E. A. Silver, & S. Li (Eds.), *Advances in mathematics education. Transforming mathematics instruction. Multiple approaches and practices* (pp. 21–36). Cham: Springer.

Weigand, H.-G., & Weth, T. (2002). *Computer im Mathematikunterricht. Neue Wege zu alten Zielen*. Heidelberg: Spektrum.

Winter, H. (1981). Der didaktische Stellenwert des Sachrechnens im Mathematikunterricht der Grund- und Hauptschule. *Pädagogische Welt*, 666–674.

Winter, H. (1994). Modelle als Konstrukte zwischen lebensweltlichen Situationen und arithmetischen Begriffen. *Grundschule, 3*, 10–13.

Winter, H. (1996). Mathematikunterricht und Allgemeinbildung. *Mitteilungen der Gesellschaft für Didaktik der Mathematik, 61*, 37–46.

Winter, H. (2004). Die Umwelt mit Zahlen erfassen: Modellbildung. In G. H. Müller, H. Steinbring, & E. C. Wittmann (Eds.), *Arithmetik als Prozess* (pp. 107–130). Seelze: Kallmeyer.

Zais, T., & Grund, K.-H. (1991). Grundpositionen zum anwendungsorientierten Mathematikunterricht bei besonderer Berücksichtigung des Modellierungsprozesses. *Der Mathematikunterricht, 37*(5), 4–17.

Zech, F. (2002). *Grundkurs Mathematikdidaktik: Theoretische und praktische Anleitungen für das Lehren und Lernen von Mathematik,* 10th edn. *Beltz Pädagogik*. Weinheim: Beltz.

Ziegenbalg, J., Ziegenbalg, O., & Ziegenbalg, B. (2010). *Algorithmen von Hammurapi bis Gödel. Mit Beispielen aus den Computeralgebra Systemen Mathematica und Maxima*. Frankfurt: Harry Deutsch.

Zöttl, L., Ufer, S., & Reiss, K. (2010). Modelling with heuristic worked examples in the KOMMA learning environment. *Journal für Mathematik-Didaktik, 31*(1), 143–165.

www.ingramcontent.com/pod-product-compliance
Lightning Source LLC
LaVergne TN
LVHW082240200926
848484LV00008B/2382
9781013267536